2nd Fully Revised and Enlarged Edition

Pests of Stored Grains and Their Management

AF631063

M.C. Bhargava

Former Dean, College of Agriculture, Lalsot, Rajasthan
Former Head, Department of Entomology, College of Agriculture
SKN Agriculture University, Jobner, Rajasthan
&
Former Principal Investigator of ICAR Project entitled
Bio-ecology and Management of Stored Grain Pests on
Different Commodities in Rajasthan

K.C. Kumawat

Profesor
Professor, Department of Entomology
SKN College of Agriculture
SKN Agricultural University
Jobner, Rajasthan

nipa®

NIPA® GENX ELECTRONIC RESOURCES & SOLUTIONS P. LTD.
New Delhi-110 034

NIPA® GENX ELECTRONIC
RESOURCES & SOLUTIONS P. LTD.

101,103, Vikas Surya Plaza, CU Block
L.S.C. Market, Pitam Pura, New Delhi-110 034
Ph : +91 11 27341616, 27341717, 27341718
E-mail: newindiapublishingagency@gmail.com
web: www.nipabooks.com

For customer assistance, please contact
Phone: + 91-11-27 34 17 17
Fax: + 91-11- 27 34 16 16
E-Mail: feedbacks@nipabooks.com

ISBN: 978-81-19254-72-9

Composed and Designed by NIPA®.

Preface to the Second Edition

The 2nd revised edition of the book "Pests of Stored Grains and Their Management" covers syllabus as recommended by the 5th Deans Committee. The book covers updated information written in simple, lucid language, easily understandable by readers and summarizes the knowledge of insects and other pests of stored grains and grain products covering global scenario.

Every chapter covers wider aspects of related work, storage requirement to prevent the losses of food grains at post harvest handling and at other levels too, different types of storage techniques and prevalent rural and improved storage structures and receptacles, storage pests (insects, mites, birds, rodents, microorganisms etc.), fumigants and their use, safety measures against poisoning, management of stored grain pests etc.

The revised edition gives the readers the vast knowledge about the progress made in different aspects of storage entomology. The book will serve as the valuable source of information on the storage entomology and would be of great importance for its readers.

The book has good number of MCQ's at the end of the book to help students along with colour images of insects and pests to easily identify them.

M.C. Bhargava
K.C. Kumawat

Contents

Chapter 1

Introduction

A series of efforts have been made to initiate the work on post-harvest grain technology in India. H.M. Lefroy, the 1st Imperial Entomologist appointed by the British Government in India in 1905 had given a tentative list of storage insects in 1906. This list of storage insects was further supplemented by T.B. Fletcher in 1914. The problems of grain storage were discussed at Pusa, Bihar at the Entomological meetings held during the years 1917, 1919 and 1923.

The Department of Food was established in 1942 by the Government of India. The executive wing of the Food Department was responsible for the purchase, movement, transport and distribution of food grains. Central Food Technology Research Institute (CFTRI) at Mysore was established in 1950 with a mandate to train personnel and evolve technology relating to food processing and preservation. The Central Warehousing Corporation was set up on 2nd March, 1957 under Agricultural Produce (Warehousing and Development) Act 1956. On 1st January, 1965 Food Corporation of India was established.

In 1958, the Department of Food started a centre at Hapur known as Grain Storage Research and Training Centre with the main objective to train the personnel of State Government engaged in food procurement, storage and distribution. This centre was upgraded during 1968-1973 with the technical and financial assistance of UNDP and this centre was redesignated as Indian Grain Storage Institute (IGSI). Two field stations at Ludhiana and Bapatla (later shifted to Hyderabad in March 1979) were established in 1968-69. Subsequently, three stations of IGSI at Jabalpur (October, 1980), Udaipur (February, 1981) and Jorhat (May, 1981) were also established. The overall objective was to minimize losses of food grains by development of modernized grain storage practices for the farmers, traders, cooperatives and large storage of food grains in the country and also to evolve the methodology to be adopted by the Save Grain Compaign which has its headquarters in the Department of Food.

In India, post harvest losses caused by unscientific storage, rodents, insects, microorganisms, moisture etc. account for about 10.0 per cent of total food grains. A world survey conducted by FAO indicated about 5.0 per cent loss of cereals in storage annually. However, conservation of reserve food grain stock is necessary to ensure an incessant supply at stable prices. Insect pests causing losses are the most serious problem in grain storage, particularly in villages and towns in developing countries, because of humid tropical condition, poor sanitation and inappropriate sanitary facilities. In the present scenario, food quality and management efficiency are the most important guidelines for planning and any decision making process, it is even more important that all people responsible for storage and conservation of agricultural products concentrate their efforts to reduce the currently high level of product loss.

Storage of food grains is repeated interim phase in the complex logistics of moving grains from producers to processors and the grain products from processors to consumers. Its purpose is to equilibrate fluctuations between two major factors, supply and demand. These two factors could be seasonal, annual, demand induced and weather induced. Due to logistic and transport delays or other causes; storage needs is a function of variability of these factors and the level of protection one desires from the effects of fluctuations. Managing store wisely with minimum loss and maintaining its quality is a major concern at all stages. How much, where and when to exercise their control depends largely on socio-economic and political conditions within and outside a country.

About 65-70 per cent of total food grains is stored by farmers in traditional bulk and bag storages for food, feed and seed purposes till the next harvest. Rest 30-35per cent is procured and stored by food grain organizations, like Food Corporation of India (FCI), Central Warehousing Corporation (CWC), State Warehousing Corporation (SWC), Cooperatives, Rural Godowns and other state government agencies.

Safe storage of agricultural produce and grains is a problem in almost all the tropical countries because of high humidity and warm climate. Storage of agricultural produce, particularly cereals, pulses and oilseeds for a long duration is associated with the losses in quality, quantity, viability and monetary values. The main bio-deteriorating agents are rodents, insects, mites and microorganisms. Insects occupy a key position among these bio-agents. About 500 species of insects have been associated with stored grain products. These are mostly coleopterans (beetles and weevils), a few species of lepidopterans (moths) and psocopterans. Nearly 100 species of insect pests of stored products cause economic losses.

Success achieved so far in making the stored products free from pests has been largely dependent on pesticides alone. Pesticides are the most powerful tools available for pest control. They are highly effective, rapid in curative action and adaptable for most situations, flexible in meeting changing agronomic and ecological conditions and relatively economical. Despite these credentials, the long and indiscriminate use of pesticides has been found ecologically unsound; associated with disadvantages like, pest resistance, out breaks of secondary pests, adverse effects on non-target animals, objectionable pesticide residues and direct hazards to the users. It has been owing to the men's tendency to substitute pesticides for effective bio-environmental and socio-economic control rather than restrict their use to the emergent situations, necessitating the pest management in the present situations.

Fortunately, India is one of the foremost countries in the world receiving inexhaustible supply of solar radiation, which receive an enormous amount of solar energy, on an average 5 kwh/ m^2 per day for over 300 days per year. The scope of harnessing the solar energy for disinfecting the stored products has ample potential in the country.

Chapter 2

Insect Infestation and Losses in Storage

Insect infestation occurs in stored grains and grain products to variable extent depending upon the storage conditions in developing countries. The losses of nutritional compounds are proportional to the extent of insect infestation, nature of insect feeding and distribution of nutritional chemicals in different seed components.

1. Sources of Insect Infestation

There are many sources that leads to infestation in storage, *viz.,* left over seed from bins or stores or spilled seeds under the stacks, cleanings, sweepings of storage structure, processing plants, old infested stores, old gunny bags, insect infested transport vehicles, storing insect infested seeds with fresh stock or *vice-versa*, entry of insects from neighbouring stores and carry over of field infestation

- Some storage insects, like Angoumois grain moth and lesser grain borer infest grains right from the standing mature crop and their infestation migrates to storage through transportation, crop drying, threshing and grain drying during post harvest operations. This infestation is termed as horizontal infestation.
- Lateral infestation of storage is commonly commenced during the post harvest operations from other sources like

previously stacked and infested crop on the threshing floor, previously used and infested transport and stored grain left from last season harvest in the same storage. This infestation is termed as cross infestation.

- After storing, insects infest grains through their latent or hidden stages from the walls cracking and surface of the storage containers or bags called as latent infestation.
- A number of Coleopteran beetles and weevils are also capable of migrating upward or downward among the grain bulk irrespective of the pressure. Thus infestation commences on the top of the bulk grains may migrate to the bottom and *vice-versa* which is known as vertical infestation.

2. Factors Responsible to Enhance Insect Infestation in Stored Grains

In storage, no doubt, a number of biotic and abiotic factors and their interactions are responsible for insect multiplication and subsequent grain damage. Major factors responsible for enhancement of insect damage are:

- High grain moisture content due to prevailing high atmospheric humidity generally above the safe moisture levels (14% for paddy, 13% for milled rice, 12% for wheat, 10% for oilseeds and 9% for pulses).
- Fewer fluctuations in diurnal and seasonal temperatures.
- Unscientific fabricated or constructed traditional storage structures which are neither moisture proof nor can prevent cross infestation of the insects.

This is the reason that storage losses are always more and occur throughout the year in the coastal states of India because of high humid and warm climate with less fluctuation in diurnal and seasonal temperature as compared to Northern and Central states, where humidity is relatively less and generally coupled with fluctuations of temperature, which are rather extreme during winter and summer months. Extreme temperatures and low humidity inhibit the population build up of the insects.

Physical environment: Local climatic conditions including temperature, humidity and grain moisture influence insect infestation in stored food grains as discussed below:

(a) *Humidity, grain moisture and rainfall*: Atmospheric humidity is directly related to the grain moisture content and positively correlated with insect infestation. Moisture present in the grain or kernels is in two forms, water of composition which is contained within the plant cells of which the grain kernel is composed of and absorbed water or free water which is present in the surface, but not within the cells. The quality of free water (absorbed water) held by the grain product is a critical factor which plays a vital role in the safe storage of food grains. The weather affects the moisture content of grain before and after harvesting during storage. The grain may absorb the moisture depending on the relative humidity of surrounding environment. Excessive moisture in the grain is an open invitation to the dangers of sprouting, fermenting, insects and microorganism attack making them discoloured, stinking, unpalatable, poisonous and unfit for human consumption. Such food grains even become unfit for cattle and poultry feed. A grain is bad conductor of heat but its flour is comparatively good conductor. Due to the diurnal and seasonal variations of temperature, moisture migrates resulting in saking and sweating of food grains in large bulks, therefore, dunnage must always be used in bagged storage to prevent sweating of bags and prevent moisture pick up from the floor.

If a grain bulk which is clean, dry and sound in every respect is allowed to come in contact with atmosphere of high relative humidity, the food grains soon adjust themselves in equilibrium with surrounding atmosphere and the increased grain moisture thus gives every chance for insects, mites and microorganisms to develop. In addition to this, the embryo which is alive, becomes active and has a tendency to germinate.

Rainfall is another variable which is related to moisture content of the grains and insect infestation. During rainy season, grain moisture of stored grains generally increased due to the increase of high atmospheric humidity because most of the storage structures or godowns are not moisture proof. High grain moisture is not only conducive to insect infestation but also activates massive growth of microorganism like storage fungi.

(b) *Temperature* : The temperature in storage structures or godowns as well as atmosphere is found to influence the insect infestation in combination with relative humidity and control the rate of metabolism, growth, development, reproduction and behaviour in general. In a storage ecosystem, temperature and relative humidity fluctuate within defined range which allows the insect to survive and multiply. Temperature ranging from 20-40°C accelerates the development and reproduction. Temperature below 10°C for prolonged period may cause death. Temperature ranging from very low about 2°C to a maximum of 63°C may influence the growth rate of storage fungi. Bacteria will develop up to a temperature of 71°C. Under very damp conditions, grain respiration increases to the point where germination occurs. The lower the temperature, the greater the retarding influence on respiration rate. The lower the initial moisture content from which artificial drying takes place, the less detrimental is the effect of high temperature on germination.

Temperature is the most important factor controlling the development and multiplication of insects in general. Besides normal temperature effects, there are number of harmful side complications arising out of improper and ineffective manipulations of temperature mainly due to hot spots developing within the stored grains. Another is the development of temperature gradient generally from the middle towards the periphery. Grains stored in storage structures have low moisture content but not so low to rule out the possibility of production of some insect species. At favourable temperature, insects reproduce rapidly to build up population at certain places in the bulk grain. The insect may respire at the rate many thousands time more than that of the same weight of the grain, thus heat is generated which rises the temperature of bulk grain considerably. Such heating is known as dry heating which is due to the combined respiration of insects and grains results in the formation of hot spots which forces the insects to migrate to cooler area.

Before putting the grains in the store, it should be well dried and cooled. If hot grains are kept in storage, it starts respiring at a faster rate and thus more heat will be generated resulting in moisture migration, creating hot pockets, discolouration of grain, stick grains and charring etc.

(c) *Oxygen*: A grain is a living organism which respires during storage. For respiration oxygen is a must, the quantity required may be small or large. In absence of oxygen the grain may die out and lose its germinating power. No doubt, lack of oxygen may help in reducing the insect infestation but destroy the viability of the grains to make it unfit for seed purpose. On the other hand, normal oxygen supply may keep the grain seed living but it provides ample facilities to insect, mite and micro-flora multiplication, hence, a restricted supply of oxygen is desirable for safe storage of food grains as well as seeds.

(d) *Storage structures and storage periods*: Local storage structures which are commonly used in rural India and other South-East Asian countries generally fail to provide complete protection from insects as these structures are not moisture proof. These storage structures do not permit disinfestation procedures effectively, like treatment of storage containers and inturn allow cross infestation. Gunny bags or jute bags are more prone to infestation of insects than bulk structures. Unhygienic conditions of storage structures are also found to influence insect infestability to food grains. Storage period is a vital factor, which affects insect infestation. In rural India, paddy is generally stored for a period of 6-9 months and only a few farmers retain their paddy in storage for a year or more. Longer the storage period, higher is the insect infestation. Stored paddy is subjected to infestation of only *S.cerealella, if* stored for 6 months and *S. cerealella, R. dominica, S. oryzae* and *T. castaneum if* stored for more than 6 months to one year.

(e) *Grain drying*: After harvest, sun drying is the most common method for removing moisture of the foodgrains as use *of* mechanical dryers is not economical at farmers level. Sun drying at high light intensity (11 AM to 2 PM) usually leads to cracking in the grains which inturn enhances insect infestation in storage.

3. Losses Caused by Stored Grain Insect Pests

Losses in storage are of multiple nature, including loss in weight, quality, nutritive and market value. Each of these types of losses may have different significance which varies with people, time, place and with existing loss assessment methods and techniques. The losses are not limited to technical and physical factors only. The socio-economic realities do play an important role, particularly in the developing countries. A precise, comprehensive and uniform method for expressing storage losses is yet not available. The process of determining weight changes has not been given due consideration for all the variants, *viz.*, moisture from atmosphere, moisture through insect metabolism, heating, moisture migration and presence of insect contaminants and socio-economic factors.

Different kinds of losses due to storage insect pest infestation in food grains are as follows:

- Quantitative or weight losses
- Qualitative losses
- Losses of seed viability
- Damage to storage containers

Quantitative or Weight Losses

Government of India Committee for estimates of losses indicated that when insect damage was quantified, losses in various food grains were 3 per cent in wheat, 2 per cent in rice, 2 per cent in *jowar*, 3 per cent in *bajra*, 5 per cent in millets, 5 per cent in gram, 5 per cent in pulses and on as average, it was 2.55 per cent. Similarly, Department of Agriculture, Government of Gujarat conducted survey in 1966 and reported 8.7 per cent losses in *bajra* and 9.84 per cent in maize. Srivastava *et al.* (1971) reported 0.2 to 3.6 per cent loss in weight due to insect damage in well-managed cooperative seed stores. Further, Srivastava *et al.* (1973) found 9.7 per cent weight loss due to insect in a farm level storage practice in wheat and kernel damage up to 30 per cent. Girish and Jain (1974) recorded 0.06 to 9.7 per cent weight loss due to feeding of insects during 6 months of wheat storage. In Punjab, Doharey *et al.* (1975) conducted a preliminary survey in some selected grain markets and observed 0.03 to 4.02 per cent storage losses due to insects alone in 10 months of storage. Lal (1984) also studied the quantitative and qualitative losses in wheat due to storage insects at farm level in Madhya Pradesh and reported only 0.06 per cent insect damage in metal bin and 1.17 per cent damage in improved *Kachchi Kothi* after 12 months of storage in a Save Grain Campaign adopted village, while in non adopted villages the minimum weight loss was 2.7 per cent in *Banda* (out door) and maximum 11.2 per cent in *Palai* after 9 months of storage.

Qualitative Losses

Apart from physical loss in weight, there are qualitative depletions in nutrient contents, caloric value and other adverse transformation during storage that must be considered

Qualitative losses caused by storage insects are categorized in four main heads as follows:

- Loss of nutritive and caloric value
- Chemical changes in grain content
- Contamination of grain with pathogenic and toxicogenic microbes
- Increase in moisture content

(i) *Loss of nutritive and caloric value :* Storage insects when feed on the kernels and destroy the germs, directly cause caloric loss. Such feeding in fact causes not only weight losses but it also deteriorates the quality of the grains when assessed in the buyer's market. The nutritional quality of the grains is adversely affected by a number of agencies. Deterioration of food grain quality initiates immediately after fertilization and embryonic development of the cotyledons by the disturbance of enzyme sequence. When storage insects feed and develop inside or on the grains, also caused bio-chemical degradation.

(ii) *Chemical changes in grain content :* Bio-chemical changes in grains take place such as changes in carbohydrates, proteins, fats and vitamins.

(iii) *Contamination of grain with pathogenic and toxicogenic microbes :* Contamination of grains with pathogenic and toxicogenic microbes by the storage insects is found to cause loss of palatability, bad odour, allergy to human beings and food poisoning as described below.

  (a) *Loss of palatability*: Contamination of food grains with body excretes, scales, fragments of bodies and dead-bodies are common when insects multiply in stored grains. Insect fragments and excrements contamination is closely related and dependent upon feeding species which leads to the loss of palatability and bad odour

(b) *Bad Odour*: The odour from excrements is observed to be musty in the case of *Rhizopertha dominica* and *Tribolium castaneum*. Grains stored at 15 per cent moisture for one year showed not only reduction in iodine number in its oil but also favoured insect infestation and mould infection which resulted in poor taste and odour.

(c) *Allergy to human beings*: Body parts of insects ingested with food grains cause formation of antibody and some times allergy in human. However, positive skin reactions are reported in group of patients by *T. castaneum*, *S. oryzae* and *R. dominica*. These insects also cause allergic reactions by contactants, injectant and inhalant allergens.

(d) *Food poisoning*: Insects infesting stored grains are known to act as carriers of several pathogenic and toxicogenic fungi and bacteria and to provide ecological conditions for better development and increased biochemical degradation. Some pathogenic and toxicogenic fungi and bacteria cause food poisoning and health hazards.

(iv) *Increase in grain moisture content :* Infestation of storage insects also enhanced the grain moisture during the course of their multiplication. Grain moisture of 14.6-14.8 per cent increased 17.6-23.0 per cent due to damage by *Sitophilus granarius* (Agrawal *et al.*, 1957). Insect infestation in stored rice is found to increase moisture content and allow moulds to grow (Kauraw and Prakash, 1980). *S.oryzae* multiplication enhanced grain moisture from 14.0 to 23.6 per cent in milled rice variety (CR-1014) during 9 months of storage under controlled conditions. In these grains, fungi also developed luxuriously and completely spoiled (Prakash *et al.*, 1987).

Losses of Seed Viability

Retention of seed viability during storage is essential to maintain crop stand in the field, ultimately to increase crop productivity. Insects attack the seed, first damage the germ which infact is much softer than rest of the kernel and as a result its germination is lost. In paddy storage, seed viability deteriorates individually as well as collectively by the interactions of insects, fungi, grain moisture and temperature. But it is difficult to estimate the viability losses due to insect damage alone, where visible damage symptoms are not clear. Grain boring insects like *S. cerealella* and *R. dominica* are also known to cause invisible loss of viability due to attack of their first instar larvae to the germ of seed and sudden death without further development. This kind of seed damage can not be seen as there is no insect emergence hole in the seed and its weight is almost remain intact and thus difficult to separate out at the time of seed soaking in water before sowing.

Germination of stored seed tends to decrease with the increase of insect infestation. Prakash and Kauraw (1982) found losses of viability ranging from 3.67 to 41 per cent in paddy seed sample possessing 1.43 to 14.05 per cent infestation of storage insect during 3-9 months of storage under warm and humid climate in Orissa. In a study, Narsimhan *et al.* (1985) reported 20-28 per cent loss of paddy seed viability due to infestation of *S. cerealella* over 6-9 months of paddy storage in the top position of bamboo weaven containers and 8-10 per cent due to *R. dominica* in jute bags during 6 months of storage.

Damage to Storage Containers

R. dominica is a notorious pest, which has the capacity to destroy wooden based storage containers and even polyethylene lined in the storage and also almost all local containers made of material other than cement and iron. This insect damage all the structures and materials by making the holes in them, which help other insects like *S. oryzae* and *Tribolium* spp to infest or penetrate through these holes inside the structures. Considerable damage to wooden framing in cargo ships is caused by *R. dominica*.

Chapter 3

Estimation and Detection of Insect Infestation in the Grains and Milled Cereal Products

The damage to food grains in the storage may be caused due to insect attack, mould, fungi, bacteria, rodents, heat and moisture. The value of grain for milling purposes and human food is directly correlated with the amount of insect contamination present. The immature stages of rice and granary weevils, lesser grain borer, Angoumois grain moth and occasionally the flat grain beetle, cadelle, rust red flour beetle and saw toothed grain beetle live inside the kernels where they are hidden from view until they emerge.

Grain from casual inspection appears to be in good condition but may contain enough internal infestation which render it unfit for human food or unfit for purchase for future sale as milling stock. In general, millers have found that wheat with more than 0.5 per cent of infested kernels is unfit for milling. The need for a rapid and accurate method of estimating insect contamination in grain offered for sale has become more and more important to both grain dealers and millers.

Methods for Estimation of Damage

In order to assess the damage caused to food grains due to any one or combination of various factors; physical, chemical and micro-analytical methods are employed according to the necessity and information required. Various methods have been employed by field workers as well as by laboratory technicians in estimating damaged grains in food grains with particular reference to detection of insect infestation which alone cause maximum damage. These methods have been described below.

(i) Physical Methods

(a) *Visual examination*: The condition of the stock, *vis-a-vis* its appearance, lustre etc. is directly affected by the presence of damaged kernels including insect infested kernels in the grains. A trained worker can very easily determine the general qualities of the grain by its appearance. The presence of damaged grains in the samples is indicated by loss in its natural lustre which appears to be rather dull and lifeless. This method is usually carried out as a preliminary to other tests.

(b) *Germ damage and insect emergence holes*: The per cent kernels damaged by insects including the germ eaten kernels can be assessed directly by physical analysis of the sample. In this method, 100 g of grain sample is evenly spread out on a glass plate; the damaged kernels are picked up to assess the percentage of insect damaged kernels. This method is very simple and gives a good index of deterioration or damage to the field workers but it does not indicate the damage going on inside the kernels which is not visible to the naked eyes.

(c) *Determination of foreign matter, insect infestation etc.*: The foreign matter and insect debris, webbing etc. are physically separated with the help of sieves and various insects, debris, larvae etc. by hand picking. The presence of insects, debris and webbings, larvae and their cast off exuviae etc. give a rough idea of the damage. The foreign matter is also examined under microscope to determine the type of infestation and damage.

(d) *Total damage*: The determination of total damage in the samples gives fairly good idea about the condition of the grains. In this case, heat damaged including sprouted, moulded and rotten wheat kernels are physically separated. This is the usual method followed in the commercial channel in determining the quality of wheat.

(e) *Germination tests*: In this method, the viability of the wheat and other food grains is reported in terms of percentage of kernels developing strong sprouts under controlled conditions by incubating a representative sample of 10 g seeds on the moist filter paper at optimum temperature. Reduced viability indicates increase in deteriorated or damaged kernels in a given sample.

(f) *Loss in weight*: Damage caused due to insect infestation directly results in loss in weight which can be estimated by determining volume weight ratio. A decrease in volume/ weight ratio indicates an increase in damaged kernels in a given sample. The per cent dry mass loss is worked out by count and weight method (Dick 1987).

$$\text{Dry mass loss (\%)} = \frac{(UNd) - (DNu)}{U\,(Nd + Nu)} \times 100$$

Where,

Nu = Number of undamaged grains.

Nd = Number of damaged grains.

U = Dry mass of undamaged grains, and

D = Dry mass of damaged grains

(ii) Organoleptic Qualities

The determination of the organoleptic qualities of samples gives a fairly good idea about the damage caused. The sample to be tested is cooked (incase of wheat) and evaluated by a panel of six judges for texture, odour, smell, taste and acceptability of the product for consumption. It has been established that the products made out of food grains containing high percentage of damaged kernels are not acceptable.

(iii) Detection of Insect Infestation

Since, the damage (direct as well as indirect) caused by insects resulted in maximum damage to the food grains, the timely detection of insect infestation is most important in minimizing the damage to food grains. The following methods are usually adopted for detecting the infestation.

a. Visual Examination

In most cases where grains contain an appreciable amount of insect infestation, there will be a large enough percentage of kernels with insect emergence holes to give a rough indication of the actual amount of infestation present. The infested sample is sieved with the help of 10-12 mesh sieve, refractions examined for living or dead insects that are present and counted. The sieved sample is also spread evenly in a plate, insects collected including larvae, if any and counted. This test is usually followed by the field men as well as in the physical analysis of the grain.

b. Hidden Infestation

By the above method, the immature forms of insects that remain within the kernels as well as those which are not visible to the naked eye cannot be seen. This is known as hidden infestation of food grains. Many methods of determining the presence of dangerous amount of internal insect infestation have been suggested from time to time.

(a) *Emergence hole counts* : The number of grains with insect emergence holes in a given sample is counted. In this method, the grain is also taken out to observe the presence of larvae. The examination under the microscope to observe the immature stages of insects not visible through the naked eye is also undertaken.

(b) *Incubation*: In this method the samples of grain to be tested are incubated under controlled conditions of temperature as well as humidity in glass jars and the number of insects emerged after a period of every 24 hours are counted for a period of 15-20 days depending upon the life cycle of the insects emerging. This method is reliable but time consuming.

(c) *Stain method* : In this method, solution of various dyes is used for staining egg plugs in a given sample. The stained insect plugs are separated and estimated. The stained kernels must be examined closely to differentiate whether the stained portion is a plug hole or endosperm of the grain. The principle objection of egg plug stains is that there is no indication of stages of development of the insect within the kernels. The following different methods employing various stains have been developed for detecting hidden infestation of insects.

(i) *Acid fuchsin*: This is rapid method of detecting the presence of weevil infestation within grain which was described by Frankenfield (1948). He employed an acid fuchsin stain whereby the gelatinous weevil egg plugs in grain were stained a bright cherry red and feeding punctures and mechanical injury stained a light pink. The stain was prepared by mixing 50.0 ml glacial acetic acid in 950 ml distilled water and adding 0.5 g acid fuchsin. Uniform samples of grains prepared by soaking in warm water for 5 minutes were immersed in the stain for 2-5 minutes after which the excess stain was removed by washing in tap water. The egg plugs are about the size of an ordinary pin prick and can readily be seen with the naked eye.

(ii) *Gentian violet*: Goosens (1949) suggested a gentian violet method for determing the hidden infestation. The wheat is exposed for 2 minutes in a solution containing 10 drops of 1 per cent aqueous stock solution of gentian violet in 50 ml of 95 per cent ethanol. The egg plugs are stained purple.

(iii) *Iodine solution*: The sample to be tested is immersed in iodine solution which gives a dark brown stain on the egg plug. While examining stained grains care must be taken in separating kernels with broken skin as such kernels are likely to be stained by this solution.

(iv) *Berberine sulphate*: A fluorescent stain berberine sulphate was suggested by Milner *et al.* (1950) for staining the egg plugs but has not come into general use. By this method, kernels of wheat are emerged in dilute solution of berberine sulphate (20 ppm) for one minute. When exposed to ultra violet light the egg plugs show an intense yellow fluorescence.

(d) *Carbon dioxide determination*: Howe and Oxley (1944) proposed a method based on the fact that insect breeding in grain produce carbon dioxide. By measuring the quantity of carbon dioxide produced by the given sample of grain in 24 hours, the extent of internal infestation could be estimated. If the carbon dioxide concentration exceeds one per cent, it is certain that a potentially dangerous infestation is present. A figure of 0.3 per cent may represent insect free grain if the moisture content of the grain is 14 per cent or above. If the moisture content is below 14 per cent, such a reading would indicate a slight infestation. A reading between 0.3-0.5 per cent indicates either a slight insect infestation or a moisture content higher than 15 per cent. A figure between 0.5 and 1.0 per cent indicates that the grain is unfit for prolonged storage. This method requires too long a period and fails to indicate the presence of dead insects.

(e) *Flotation method*: Studies by White (1956) indicate that a mixture of two solutions of different specific gravity can be used to make a rapid separation of infested from noninfested kernels in a sample of wheat. The mixture consists of a solution of sodium silicate in water with a specific gravity of 1.60 to which is added methyl chloroform adjusted to a specific gravity of 1.30 with deobase oil. When placed together, a definite separation layer is formed between the two liquids. The lighter sodium silicate solution remains on top of the heavier methyl chloroform-oil solutions. Kernels containing the later stages of weevil larvae float on top of the sodium silicate solutions. Kernels containing early stages of weevil larvae and some noninfested light weight kernels are buoyed up to the separation level of the two liquids by the 1.30 specific gravity methyl chloroform-oil mixture. Noninfested, normal weight kernels sink to the bottom. By placing 1000 kernels sample in a beaker of the mixture and stirring it to wet the kernels, a quick separation of the kernels occurs. If there are no kernels in the top layer, the sample can be considered free from serious infestation. If the presence of floaters; the degree of infestation can be estimated by the relation between the number of floaters and the size of the sample. This method should be of considerable value to the country elevator operators who wish a rapid and reasonably accurate method of determining the condition of grain offered for sale.

Apt (1952) suggested a floatation technique whereby a 100 g sample of wheat is placed in a 2 per cent ferric nitrate solution, made by adding 2 g of hydrated ferric nitrate to 100 ml of water. After 30 seconds of agitation of the samples to wet the kernels, those with weevil emergence holes will float and can be counted.

(f) *Sodium hydroxide gelatinization method*: This method was suggested by Apt (1950) for quickly detecting the presence of insect infestation in wheat. The treatment consists of boiling the grain for ten minutes in a 10 per cent solution of sodium hydroxide. This method has been adopted on a large scale.

(g) *Cracking flotation method:* One of the most accurate methods of determining the amount of internal insect infestation in grains is known as 'cracking flotation method'. Cleaned grain is coarsely ground to release the internal insects and is either soaked in water alcohol mixture or in boiling water then mixed with gasoline or mineral oil. The insects are floated off with the oil layer in a Wildman trap flask and collected on a filter paper and counted. In the process of grinding, some of the insects are broken, but characteristic parts such as the heads can be counted and a fairly reliable estimate of the number of whole insects arrived at. For domestic flour production, some mills will not use wheat for their mill mixes containing more than eight internal insects per 100 g. A detailed description of the 'cracking floatation method' used by Harris *et al.* (1952) is as follows:

After cutting out slightly over 100 g with the Jones sampler, weigh out 100 g. Transfer a small amount at a time on to a 5 or 8 inch No. 12 U.S. standard sieve and brush with a stiff bristle brush to work the surface insects through the sieve as nearly quantitatively as practicable. Grind the screened wheat in a burr or Wiley mill set just fine enough to cut the kernels approximately eights or with no granules may coarser than 2 mm. Transfer all of the cracked wheat to a 2 litres trap flask and brush into the container any residue that may remain in the mill after grinding. Add about 400 ml of 60 per cent alcohol and mix thoroughly. Wash down the sides of the flasks with sufficient 60 per cent alcohol to bring the total volume to 700-900 ml and soak for 10 minutes. Add 35-50 ml of gasoline, mix thoroughly and allow to stand for 30 minutes and trap off into a beaker. Mix 25-30 ml of gasoline into the trap flask, then 5 ml of 60 per cent alcohol and after 30 minutes trap off into the beaker. Filter through a 10 XX bolting cloth or if a starchy residue remains add sufficient concentrated HCl to make 1-2 per cent acid, bring to a boil, then filter through a 10°C XX bolting cloth and examine at 15-30°C X. Count only whole adults, larva or pupae or the heads, head capsules or cast off skins from adults, larvae or pupae. This method, while reasonable accurate, is time consuming

(h) *X- ray method*: This is most accurate and rapid method of determining internal insect infestation and has the advantage that all stages of development of the insects can be observed. This method was described by Milner *et al.* (1950). In this method, X-ray machines are used to take radiographs of 100 g sample of wheat. These radiographs reveal the presence of insect forms within the kernels. X- ray manufacturers have developed X- ray units specifically for this purpose. Equipment is rather expensive but is being used extensively by large milling concerns.

(i) *Ninhydrin process*: Dennis and Decker (1962) suggested the use of a chemical indicator technique for determining the presence of living insect infestation in grains, based on a colour reaction between the body fluids with ninhydrin-impregnated filter paper. In a machine developed for the purpose, grain to be tested is fed into the machine into a fold of filter paper impregnated with a 0.7% ninhydrine acetone solution. The tape containing the kernels is passed through rolls which crush the kernels, allowing the body fluids of insects within the kernels to react with the chemical in the filter paper forming purple spots which can be counted.

(j) *Detection of insect fragments*: The insect fragments observed in a given sample of grain can be determined by grinding the material and placing the same in a clove oil base under microscope.

(k) *Maltose test*: An abnormally high maltose test also indicates deteriorative conditions of the grain. In this test, the amount of reducing sugar is measured. The amount of reducing sugar is determined with thiosulphate and results expressed as mg in 100 g flour.

(l) *Fat acidity*: Deterioration of grain and milled products in storage is accompanied by an increase in fat acidity. Zeleny (1954) has reported a correlation between the fat acidity and percentage of damaged kernels present in a given sample of grain.

(m) *Detection of rodent contamination*: For the detection of fragments of rodents pellets and urine contaminant in grain, analytical procedures have to be followed. These methods essentially consist of separating the pellet fragments by density separation method in a chloroform and carbon tetrachloride mixture and examining the residues under microscope. This, however, requires the experience and training of an expert. Rodent hairs are also separated and examined microscopically in the same sample. For testing urine spots, Buckley Whinerey (1945) prescribed a simple test in which by applying a solution of paradiumethyl amino benzal dehyde, ethyl alcohol and oxalic acid, a brown yellow colour is obtained, if urine stains on the surface of the grain also fluorescents when subjected to ultra-violet light and indicate a white- greenish light.

(n) *Temperature and moisture*: The grain is a living organism and like any other organisms, moisture temperature as well as the condition under which it is stored including insect infestations are important factors that lead to the deterioration in storage. It has been demonstrated that unsound grain with a high percentage of damaged kernels harbour greater number of microorganisms, insects etc. and high moisture content is more likely to cause heating in storage than the sound grain. The heating is due to respiration of the grain as well as microflora of insect infestations under circumstances. Thus, regular determination of the temperature and moisture content will help in assessing the true condition of the grain. Any spontaneous rise in the temperature or moisture content in a particular area of the grain usually indicates the presence of damaged grain or insect infestation or both and is a positive sign of deterioration. The determination of the temperature and moisture content greatly helps in knowing the condition of the grain time to time. This practice is regularly followed as a routine in grain storage. The temperature can be measured by feel and smell method, thermometer and thermocouple method; whereas, the moisture content of a given sample of food grains can be determined by feel and bite, air oven or electric conductivity method.

Chapter 4

Insect Pests of Stored Grains and Grain Products

A number of insects infest seeds in storage. These insects infest seeds to fulfill their feed and shelter requirements, which result into qualitative as well as quantitative loss of seeds. Indian climate is very favourable for their multiplication and growth throughout the year. Storage pests move along with commercial products from one area to another and even to different countries in the world. Some adult insects are strong fliers and may spread from one to another place of storage or from source of infestation. Generally, all insects infesting seeds in storage belongs to insect order Coleoptera and Lepidoptera (Table 4.1). Some minor species belong to Pscoptera. These pests cause losses to man by destroying large quantities of grain, corn, legumes (beans, lentils and peas), milled cereal products, flour, bran, macaroni and other paste products, dried fruits, dried vegetables, cheese, nuts, candy and other food materials. They contaminate and thus eliminate for human consumption, far more food than they eat and either cause direct loss or transmit diseases to man and his domestic animals. These insects can be categorized as major or minor pests according to the loss of seeds caused by them. Similarly, on the basis of their feeding behavior, they can be grouped as follows:

(a) *External feeder :* This group of insects feed on germ and endosperm from outside. They may attack whole seed and damage the germinal portion or feed on the seeds, which has already been damaged or infested by other insects or broken mechanically. These insects and their different stages are generally visible among the seeds, *e.g.* khapra beetle, Indian meal moth, red rust flour beetle, saw toothed beetle etc.

(b) *Internal feeder :* These insects mostly lay eggs inside or on the seeds and spent a part or entire larval and pupal life inside the seed, only emerges as adults. They contribute significant loss of germination which is not detectable outside, *e.g.* rice weevil, pulse beetle, granary weevil, Angoumois grain moth etc.

(c) *Primary storage pests :* They can damage sound grains, *e.g.* rice weevil, lesser grain borer, khapra beetle, pulse beetle, Angoumois grain moth, rice moth, Indian meal moth, almond or fig moth.

(d) *Secondary storage pest :* They damage broken or already damaged grains *e.g.* rust red flour beetle, flat grain beetle, saw-toothed grain beetle etc.

Table 4.1: Common insect pests of stored commodities

S.No.	Order/family	Scientific Name	Common Name
1	Anobidae	*Lasioderma serricome* (Fab).	Cigarette beetle
		Stegobium paniceum (Linn.)	Drug store beetle
2.	Anthribidae	*Araecerus fasciculatus* (De Geer)	Coffee bean weevil
3.	Bostrychidae	*Rhizopertha dominica* (Fab.)	Lasser grain borer
4.	Bruchidae	*Callosobruchus maculates* (Fab.)	Pulse beetle
		C. analis (Linn.)	-do-
		C. chinesis (Linn)	-do-
		Acanthoscelides obtectus (Say)	Bean beetle
		Caryedon serrutus (Olivier)	Tamarind beetle or groundnut beetle
5.	Curculionidae	*Sitohilus granaries* (Linn.)	Granary weevil
		S. oryzae (Linn.)	Rice weevil
		S. zeamais Motsch.	Maize weevil
6.	Cucujidae	*Cryptolestes acutangulus* Gyll.	Acute angled fungus beetle
		C. ferrugineus (Stephens)	Rusty grain beetle
		C. pusillus (Schoenherr)	Flat grain beetle
7.	Dermestidae	*Trogoderma granarium* Everts	Khapra beetle
		Anthrenus scrophulariae (Linn.)	Carpet beetle
		Attagenus piceus Oliv	Black carpet beetle

(Cont.)

S.No.	Order/family	Scientific Name	Common Name
8.	Ptinidae	*Gibbium psylloides* (de Czenpinski)	Hum p spider beetle
9.	Tenebrionidae	*Tenebrio molitor* Linn.	Yellow meal worm
		Triboliumcastaneum Herbst	Rust red flour beetle
		Tribolium confusum Jacq.	Confused flour beetle
		Triboluum destructor Uython.	Large flour beetle
		Latheticus oryzae Waterhouse	Long headed flour beetle
		Alphitobius laevigatus (Fab.)	Black fungus beetle
		Alphiophagus bifasciatus Medford	Two banded fungus beetle cadelle beetle
10.	Trogostidae	*Tenebroides mauritanicus* (Linn.)	Cadelle beetle
Lepidoptera			
11.	Gelechiidae	*Phthorimaea operculella* (Zell.)	Potato tuber moth
		Pectinophora gossypiella (Saunders)	Pink boll worm
		Sitotroga cerealella (Olivier)	Angoumois grain moth
12.	Galleriidae	*Aphomia gularis* (Zell.)	Stored nut moth
		Corcyra cephalanica (Staint.)	Rice moth
		Galleria mellonella Linn.	Greater wax moth
13.	Phycitidae	*Cadra cautella* (Wlk.)	Fig/almond moth
		Ephestia elutella (Hubner)	Tobacco moth
		Plodia interpunctella (Hubner)	Indian meal moth
14.	Pyralidae	*Pyralis farinalis* Linn.	Grain snout moth
15.	Tineidae	*Nemapogon granella* (Linn.)	European grain moth
		Tinea pellionella (Linn.)	Case making cloth moth
		Tineola bisselliella (Hum.)	Webbing cloth moth
		Trichophaga tapetzella (Linn.)	Carpet moth
16.	Psocoptera	*Liposcelis divinitorius* Muller	Grain lice
		L. transvallensis (Endernein)	

Cigarette beetle : *Lasioderma serricorne* (Fab.)

Coleoptera : Anobiidae

Distribution: It was originally named as *Ptinus serricorne* by Fabricius in 1792 and distributed particularly all over the world.

Identification: The adult is small robust beetle, reddish yellow in colour and about 2.5 mm in length. The body is covered with fine hair, head hidden under domed neck shield, wing without markings and antennae serrated with no club. The larvae are densely hairy and grow to a length of about 4 mm.

Hosts: It is a major pest of tobacco, both in the leaf and in cigars and cigarettes. It is also a major pest of spices, dry fruits, cocoa, various cereals, textiles and many other products. It is also secondary pest of pulses.

Nature of damage: As the name indicates, it is a pest of stored tobacco but also causes considerable damage to other stored products. Both the grubs and beetles attack, making holes through the food materials. In severe infestation, numerous small light coloured brown headed grubs or brown beetles can be seen causing damage.

Habits and life history: Beetles are strong fliers and attracted to light at night. The pest can be found throughout the year. The beetles live from 2 to 4 weeks and do not feed. After a day of emergence, female starts laying eggs in all kind of rather hidden safety niches like the space between the folds and the leaves, in nooks and corners, in crevices and depressions, in between grains like coriander etc. The eggs are ovoid, elliptical, whitish become opaque and dull in colour just before hatching. A single female lays about 110 eggs which hatch in 5 to 6 days and larval stage lasts for 25 days followed by pupal period of 5 to 7 days. Newly hatched larva is less than 1.0 mm in length and covered with fine hair. Head of larva is yellowish, body semi transparent and whitish, making small galleries in the host causing damage. After 25 to 30 days of larval life, it constructs a smooth lined cell in which it rests. Larvae lie quiescent in curved position within the cell, which gradually took pupal form. Newly formed pupa is glossy but changes to reddish brown after a few days. The pupal period varies from 5 to 31 days. Female adult is slightly bigger than males. There are generally 3 to 6 generations in a year. The developmental period is affected by both humidity and temperature. At 30°C, the period required for the complete life cycle varies from a mean of 33.9 days at 70 per cent relative humidity to 78.8 days at 40 per cent relative humidity, and that natural mortality increased from 50 to 73 per cent, respectively. At 30 per cent relative humidity, there was cent per cent mortality. The minimal temperature for development was about 18 °C and food stored below this temperature was found to be safe from infestation.

Parasites and predators : Many parasitic Hymenoptera have been associated with *L. serricorne* including *Lariophagus distinguendus*

(Foerster), *Choetospila elegans* Westwood and *Anisopteromalus calandrae* (Howard), *Cephalonomia gallicola* Ashmead. Some strains of pathogenic bacteria, *Bacillus thuringiensis* and *B. cereus* are effective against the larvae. The predatory mites *Pyemotes ventricosus* (Newport), *P. tritici* L. and *Cheyletus eruditus* (Schrank) attack younger larvae and eggs and the cadella beetle, *Tenebroides mauritanicus* (L.) attack all developmental stages.

Drug store beetle : *Stegobium paniceum* (Fab.)

Coleoptera : Anobiidae

Distribution : It was originally named as *Sitodrepa panicae* by Linnaeus in 1758. It is distributed in tropical, sub-tropical and temperate regions.

Hosts : It is particularly noted as a pest of pharmacies where it infests drugs. It also attacks leather, textile, tobacco, spices, breakfast foods, biscuits and herbs.

Identification : The beetles are cylindrical, 2.25 to 3.5 mm long and uniform brown to reddish brown. They have longitudinal rows of fine hair on the elytra. Drug store beetle is similar in appearance to the cigarette beetle; however, two physical characters can be used to differentiate them. The antennae of the cigarette beetle are serrated (like the teeth on a saw), while the antennae of the drug store beetle do not end in a 3-segmented club. The other difference is that the elytra of the drug store beetle have rows of pits giving them a striated appearance while those of the cigarette beetle are smooth.

Nature of damage : The drug store beetle attacks such a wide variety of foods and material that one anonymous quote states that it eats anything but except cast iron. It gets its name from its habit of feeding on prescription drugs. It feeds on flours, dry fruits, breads, cookies, chocolates, sweets and spices. It also feeds on non-food materials such as wool, hair, leather, horn and museum specimens. It is found in pigeon nests and is known to bore into books, wooden objects and in some cases, tin or aluminum foil and lead sheets. Larval feeding accounts for the greatest amount of damage. Museum and herbarium specimens are vulnerable to attack. Slight damage and contamination can ruin these valuable and precious items. Drug store beetles harbour symbiotic yeasts that produce vitamin B. The yeasts are deposited on the eggs as they pass through the oviduct and are consumed by the larvae during egg hatch. These yeasts enable the drug store beetle to feed and survive on many foods and other items of poor nutritional quality

Habits and life history : Female lays up to 75 eggs in the food or substrate. The larval period ranges from 4 to 20 weeks. Larva tunnels through the substrate and when fully grown build a cocoon and pupate. Pupation takes from 12 to 18 days. Adult females live approximately 13 to 65 days. The entire life cycle is generally less than two months but can be as long as seven months. The duration of the life cycle is highly dependent on the temperature and food source. Development occurs between 15 to 34°C but is optimal at about 30°C and 60 to 90 per cent relative humidity.

Parasites and predators : A number of chalcid wasps including *Entedon* sp. (Eulophidae), *Dimachus discolor* Walk, *Lariophagus distinguendus* (Pteromalidae) and the bethyloid wasp, *Cephalonomia gallicola* parasitize *S. paniceum*. This pest is also vulnerable to attack by predatory mites and beetles during development.

Coffee bean weevil: *Araecerus fasciculatus* (De Geer)

Coleoptera : Anthribidae

Distribution : It is a cosmopolitan pest of coffee, cocoa beans and spices. It was first reported from Australia by Mac Leay (1837).

Hosts : Coffee beans, cocoa beans, cardamom, ginger, turmeric, garlic, nutmeg, onion, cotton seed etc.

Identification : Adult is a robust, oval shaped dark brown beetles, similar to bruchid, 3.0-5.0 mm long, elytra with whitish setae, antennae elongate clavate. Larvae are yellowish in colour, curved and legless.

Nature of damage : It attacks the maturing crops in the field and is often brought in the store at harvest where it completes its development. It is considered a potential pest of coffee causing very high economic losses.

Habits and life History: Whitish eggs are laid on the grain near the germ. Fecundity is 130-140 eggs. The life cycle in coffee beans is comleted in 46 to 66 days. The optimum temperature and humidity for growth and development are about 30°C and 80-90 per cent, respectively.

Parasites and predators : The parasitoids *Aximopsis javensis* Girault, *A. tephrosiae* Girault, *Eurytoma tylodermatis* Ashmead, *Ptinobius texanus* Crawford and *Anisopteromalus calandrae* (Howard) are known to attack larvae or pupae and the predatory mite, *Pyemotes tritici* attacks eggs and young larvae.

Lesser grain borer : *Rhizopertha dominica* (Fab.)

Coleoptera : Bostrychidae

Distribution : Like khapra beetle, this pest is also native of Indian sub-continent but now it is most prominent in the United States, Southern Canada, Argentina and Australia.

Identification : The adult is a slender, cylindrical beetle and measures about 2.5 mm long and 0.7 mm wide. The head is globular and deflexed downwards to such an extent that it is almost hidden in a dorsal view. The body is dark brown or black in colour. There is a prominent constriction between prothorax and elytra. Adults are powerful fliers flying from godown to godown. The larva is about 3 mm long, dirty white with a light brown head and a constricted elongated body.

Host : Wheat, rice, maize, sorghum, barley, groundnut, dry fruits etc.

Nature of damage : The grubs and adults cause damage by feeding inside the grain and reducing them to mere shells with many irregular holes. When the infestation is severe, the adults produce a considerable amount of frass, spoiling more than what they eat. The flour so produced, serves as nourishment for the young larvae until they are ready to bore into the grain.

Habits and life history : The insect breeds from March to November. It hibernates in winter in the adult or larval stage. At the end of March, the adults appear, copulate and females start to lay eggs. Polyandry and polygamy are common in this insect. A single female lays 300 to 500 eggs at the rate of 4-25 eggs per day. The eggs are laid singly or in small cluster glued to the grain, on gunny bags or in cracks and crevices of the godown. They measure about 0.7 mm in length and about 0.2 mm in diameter and are almost cylindrical and rounded at one end and somewhat pointed at the other. The incubation period varies from 5 to 6 days in summer and from 9 to 11 days or even longer in winter. The newly hatched larva is straight and active and burrows into the grain and completes the remaining life cycle inside the grain. Only the first instar larvae can bore inside the damaged grains, whereas, the later instars can not due to their curved shape and swollen anterior, and remain outside the grain feeding on the starchy material. The larva moults 4-5 times and is full fed in 40 to 50 days. A full grown larva is 2 to 3 mm long, dirty white in colour with a light brown head. Generally, one larva is found in a single wheat or rice kernel but 2 to 3 may be found in maize kernel. The pupal stage, like the larval stage, is passed within the grain or in the grain dust and lasts for 7 to 10 days. The males have a longer life span than the females. There are generally 5 generations in a year.

Parasites and predators : The pteromalid wasp, *Lariophagus distinguendus*, *Choetospila elegans* and *Anisopteromalus calandrae* can parasitise larvae while adults can harbour microsporidian pathogens such as *Nosema* spp. The predatory mites, *Cheyletus* spp. and *Acarophenax* spp. prey on eggs. In a recent study, *C. elegans* was demonstrated to have the potential bio-control agents for *R. dominica* at 25°C but not at 32°C.

Pulse Beetle : *Callosobruchus chinensis* (Linn.)

C. maculatus (Fab.)
C. analis (Fab.)
C. phaseoli (Gyll.)
Coleoptera : Bruchidae

Distribution : It was first described from China in 1758. The distribution of pulse beetle is comparatively limited. So far, it has been recorded from Germany, Myanmar, Rhodesia and India.

Identification **:** The larva is recognized by its creamy white, oval and flabby body. The distinguishing characters mentioned below against their names can identify the different species :

- *C. chinensis* : The adult beetle measuring 3-4 mm in length, oval, chocolate or reddish brown and has long serrated antennae. There is a pair of white elongated prominences in the middle of the hind margin of thorax, a spine on each of the inner and outer edges of the end of hind femur and truncate elytra, not covering the pygidium.

- *C. maculatus* : The adult is 3-4.5 mm long, brown elytra with large black round blackish markings on other area, hind femora with two ridges on inner surface.

- *C. analis* : The adult is oval, some what smaller than *C. chinensis*. The female is of chocolate colour with a black trapezoidal area surrounded by white streaks on three sides. This area is located in the centre of each elytra towards the outer margin. The exposed pygidium is black with a central longitudinal white streak. The male is uniformly chocolate with a tinge of straw.

- *C. phaseoli* : Pale white longitudinal markings on the elytra.

Hosts : Moong, gram, pea, cowpea, pigeonpea, bean, black gram, lentil and other pulses. It also attacks cotton seeds, sorghum and maize.

Nature of damage : From the damage point of view the pulse beetles can be divided into three groups :

- Those attacking both in fields and in stores, e.g. *C. chinensis* and *C. phaseoli.*
- Those attacking only in stores, e.g. *C. analis.*
- Those infesting only in field, e.g. *Bruchus pisorum* and *B. affinis.*

The damage is mainly caused by the larvae which bore the kernel and feed on the inner contents. The damage is at its peak from April to September and is considerably reduced in October-November. Damage to the pulses infested with this insect is very high and quite often each and every grain is infested so the pulses become unfit for human consumption. Infested grain is often converted by the traders into flour which has a characteristic off-flavour and should not be marketed.

Habits and life history : The pest breeds actively from March to the end of November. It hibernates in winter in the larval stage. At the end of March, the adults appear and copulate immediately after emergence. A day after, the female starts laying small, oval, scale like eggs which are glued to the grain. In *C. chinensis,* more than one eggs may be laid on a grain. Thus, two or three larvae (up to 8 have been reported on a single grain) may develop in separate chambers. In the allied species, *C. analis* (predominantly on green gram), only one egg is laid on one grain. A single female of *C. chinensis* may lay 34-113 eggs at the rate of 1-37 per day. The highest egg production is in May and October and the least in April, June, July and December. The incubation period is 7-14 days in April, 4-6 days in September and 8-16 days in November. The viability of the eggs varies from 3.6 per cent in May to 76.9 per cent in August and September. The young larva bores into the grain and completes its development inside. The larval stage is completed in 10-12 days in August and September and 26-38 days in November. The hibernating larvae take 117-168 days to complete their development. The full grown larva migrates towards the periphery and comes to lie next to the seed coat where it turns into an oval white pupa. The pupal stage lasts 4-28 days depending upon the climatic conditions. The adult escapes by cutting a circular hole in the seed coat and such grains can be spotted easily. The average life span of an adult is 5-20 days. A preponderance of males occurs through out the active season. The insect passes through 7-8 overlapping generations in a year.

Parasites and predators : Larvae are attacked by the pteromalid wasp, *Choetospila elegans* and *Anisopteromalus calandrae,* and eggs by predatory mites, *Pyemotes ventricosus*.

Groundnut or tamarind beetle: *Caryedon serratus* (Olivier)

Coleoptera: Bruchidae

Distribution: In India, it has been reported from South India, Rajasthan, Gujarat, Madhya Pradesh, Maharashtra, Uttar Pradesh, Punjab, Haryana, Jammu and Kashmir, Himachal Pradesh, Tamil Nadu and Kerala. At present, it has become a serious pest of stored groundnut wherever groundnut is grown in India.

Identification: The adult beetle is reddish brown to reddish yellow in colour with small black markings on the elytra. Antenna is serrate type, 11 segmented and longer in male than female. Metathoracic femur is well developed with two rows of spines. The pygidium is exposed in case of female, while concealed in male. It has large prominent eyes and readily distinguished from the other storage pests by its very broad hind femur, which bears a conspicuous comb of spines. Larvae are white but turn to pink just before pupation. The abdomen is ten segmented with well defined uromears.

Hosts: Groundnut and taramind

Nature of damage: Both intact pods and kernels of groundnut are damaged by C. *serratus* which is the only primary pest of stored groundnut. It has been reported as a serious pest on pods and seeds of *Acacia nilotica* in the Thar desert. The insect remain in the field during the hottest months of May and June and has also been reported to infest the pods and seeds of *Prosopis cineraria* and P. *juliflora*. These three tree species thus act as the secondary host for the population build up of *C. serratus* and from them it spreads to its primary host, the groundnut.

The free fatty acid value of the oil increased with the increase in the infestation of this pest in the groundnut. The per cent loss in the protein content varied from 0.20-8.76 per cent in different varieties of groundnut. The loss in oil content varied from 1.98-6.54 per cent with increase in feeding intensity.

Habits and life history: The adults of this pest avoid light; mating took place at dusk and oviposition in darkness. The eggs are laid singly on the pods as well as on the seeds of its host or it may stick to the depressions on the shell of the nut or on the smooth testa of the kernels. The incubation period ranges from 6-15 days. The larvae are white but turn to pink just before pupation. The total larval period varies from 11-49 days which depends on the temperature and quality of food. The pupation takes place in the cocoons and period varies from 13-23 days. The total developmental period varies from 32-142 days. The pre-oviposition period is generally short and varies from 1-5 days. The oviposition period lasted from 16-19 days. The longevity of the adults is found to vary with climatic conditions and it is reported up to 150 days.

Parasites and predators: The Trichogrammatid, *Uscana caryedoni* is reported to be the important egg parasite, whereas; the *Anisopteromalus calandrae* is important pupal parasite.

Rice weevil : *Sitophilus oryzae* (Linn.)

Granary weevil, *S. granarius* (Linn.)

Coleoptera: Curculionidae

Distribution: The pest is worldwide and found practically throughout India.

Identification : The adult is a tiny, reddish brown, cylindrical weevil; about 3.5 mm long with a snout of about one third of the length of body and projecting downwards. Its elytra bear four prominent yellow brown or light reddish patches and thorax is fitted with round depression. The full grown larva is 5 mm in length and is plump, fleshy, legless creature having a white body and a yellowish brown head.

An allied species, granary weevil (*S. granarius*) is slightly larger, 4.5 mm long in size when reared on the same food. The elytra are without patches and uniformly brown and polished and the thorax has less dense elongated pits in lines. It is a poor flier. It is commonly found in temperate region.

Host : Rice, wheat, maize, jowar, barley, other food grains and processed food.

Nature of damage : The damage is done by both grubs and adults by breaking and hollowing the grains. The grubs are more destructive than the adults. It is known to attack in field too. In case of serious infestation only fragments of the pericarp with frass are left behind. The grains become unfit for human consumption and also for seed purpose. The damage caused by this pest is both quantitative and qualitative. In Northern India, the greatest damage is caused from July to November. In India, the loss to stored grains caused by this pest alone is valued at many crores of rupees every year.

Habits and life history : Female starts laying eggs after 4-5 days of emergence as adults and each individual lays 300-400 eggs. It makes a cavity and lays translucent white eggs and then plugs the hole with secreted material from the ovipositor. Eggs are laid singly and measures 0.7 mm long and 0.3 mm broad. Under optimum conditions eggs hatch in about 4 to 5 days. The young tiny legless grub bores into grain kernel. The larval stage lasts for about 19-34 days and pupal period for 3-6 days. At 70 per cent relative humidity and 28 ± 1 °C temperature, life cycle is completed within a month. The adult lives for 3-5 months. The most favourable temperatures for development are 27 to 30 °C. There are generally 5 generations in a year.

Parasites and predators : The pteromalid wasps, *Zatropus incertus*, *Lariophagus distinguendus*, *Choetospila elegans* and *Anisopteromalus calandrae* have been isolated from *S. oryzae.*

Rusty grain beetle : *Cryptolestes ferrugineus* (Stephens)

Coleoptera : Cucujidae

Distribution : This is usually cosmopolitan and found in moist tropical climates.

Hosts : Cereals and oil seeds.

Identification : Adult is 1.5-2.0 mm in length; light to dark reddish brown in colour, filiform antennae which is little more than half the size of body.

Nature of damage : Both adults and larvae feed on stored products and are important pests of mills. Larvae feed on germ portions and even on dead insects. Adults are only scavengers, cause heating in grain and flour in heavy infestation. The temperature may reach as high as 49°C.

Habits and life history **:** White eggs are laid loosely in flour, grain or crevices; incubation period being approximately 5 days. Young larva is cigar like, 0.7 mm long, and yellowish white with two reddish brown spines at anal segment. It either remains in germ of grain or moves actively. Larval period at 35ºC and 75 per cent relative humidity is about 21 days. Matured larva pupates in a gelatinous cocoon covered with dust particles. Tiny pallid young beetle emerges out from cocoon. The longevity of adult is about 6-12 months depending on the climatic conditions. The developmental period (egg to adult emergence) takes 6 weeks at optimum temperature of 28ºC and 70 per cent relative humidity.

Parasites and predators: The bethylid wasp, *Cephalonomia waterstoni* Gahan is a frequent parasite infesting the larval and pupal stages. Larvae and adults may harbour pathenogenic schizogregarines such as *Farinocystis tribolii* and *Mattesia dispora* in their hindgut.

Flat Grain Beetle : *Cryptolestes pusillus* (Schoenherr)

Coleoptera : Cucujidae

Ditribution : This beetle is widely distributed, and is often found together with *Cryptolestes ferrugineus* in moist tropical climate.

Hosts : Cereals, oil seeds, milled cereals, debris of cocoa and cowpea.

Identification : The small, flattened, reddish brown adult is only about 2 mm long. In males, antennae are not quite as long as its body, and those of the female are about half as long. The elytra have 5 parallel ridges. The beetles can both jump and fly. From a sample infested by different insects, flat grain beetles are the first to move out.

Nature of damage : It is one of the most common pests of stored grain but it cannot attack sound and uninjured kernels. The larvae also feed on dead insects and also subsist as scavengers.

Habits and life history : The female oviposits in crevices in the grain or in loose farinaceous material. The larva is slender and pale with a black head and has a pair of slender, black, spine like process at its posterior end. It is particularly fond of the germ of the wheat and many kernels are left uninjured except for the removal of the germ. The full-grown larva forms a cocoon of a gelatinous substance to which food particles adhere. Development from egg to adult may require only 5 weeks under favourable conditions, although the average in summer is about 9 weeks. The adult female may live a year.

Parasites and predators : The gregarine parasite, *Mattesia dispora.*

Khapra beetle : *Trogoderma granarium* (Everts)

Coleoptera : Dermestidae

Distribution : The native place of Khapra beetle is India but now the pest is distributed in Europe, Australia, North America and other countries of Indian sub-continent. In India, it is more confined to extreme dry climates like Punjab, Haryana, UP, Rajasthan, being less common in coastal areas.

Identification : The adults are oblong-oval beetles, approximately 1.6 to 3.0 mm long and 0.9 to 1.7 mm wide having no distinction of head, thorax and abdomen when observed dorsally. The body is stoutly built, light chocolate to dark brown or black with glabrous dorsal surface and have a retractile head. The antennae have a club of three to five segments which fit into a groove in the sides of the pronotum. . Sexual dimorphism is well developed, males being smaller in size than females. Larvae are straw coloured and hairy with dark brown hairy bands on each body segment and tuff forming tail at posterior end.

Host : Wheat, maize, rice, jowar, barley, bajra, oat, pulses etc.

Nature of damage : Only larval stage is destructive, adults being harmless. Being a primary pest, it damages the grain starting with germ portion and devouring of grain. This causes great economic loss both in quality and quantity. It is worthy to note that grubs damage the superficial layers only in bulk storage and do not burrow into layers deeper than 30-50 cm. The greatest damage is caused from July to October. The germination of seed is adversely affected.

Habits and life history : Its habit of congregation in cracks and crevices of bricks, masonry and wood storage has given it the name Khapra. It prefers low humidity and high temperature. Depending on temperature, an adult female begins to oviposit within three days of emergence and completes after 3-12 days, laying 100 to 150 translucent white eggs, singly or in groups of two or three glued to the grains indiscriminately or crevices. The eggs are milky white, turning pale yellowish with age, cylindrical, 0.7 x 0.25 mm, one end rounded, the other pointed and bearing spine like projections. At 35ºC temperature, incubation period is 4-6 days but it can vary depending upon the temperature and humidity. First stage larvae feed on broken grains and debris resulting from the feeding of older larvae as it cannot attack the whole grain. Larval period extends from 20 to 25 days at 30ºC temperature, whereas pupal life lasts for 4-8 days. It is highly resistant to starvation. Adults are short lived, *i.e.*, for about 14 days and are harmless.

Parasites and predators : The pteromalid wasp, *Choetospila elegans* is known to attack larvae. Predatory mites, *Pyemotes ventricosus* feeds on its eggs.

Black carpet beetle / *Attagenus piceus* (Olivier)

Coleoptera : Dermestide

Distribution : The pest is cosmopolitan in distribution.

Identification : The adult is 3-5 mm in length. Unlike species of *Trogoderma* and *Anthrenus*, the antennal grooves are lacking. The basal segment of the hind tarsi much shorter than the second and the middle coxae narrowly separated. The mature larva is 7-8 mm long with brownish yellow setae, short to elongate, lanceolate, and faintly ridged longitudinally. A long tuft of setae arising at random on the ninth abdominal segment is present.

Host : Cereal grains, grain products, fur, wool etc.

Nature of damage : This pest feeds on rugs, textiles including silk, carpets, processed cereals and also on matericals of animal origin such as wool, feathers, fur and fish meal.

Habits and Life History : The eggs are laid upto a total number of 60-100 per female depending on temperature and other abiotic facters. The larvae often found in abundance beeding in cracks in the floor of ware-houses where food stuffs have accumulated. In the spring, these larvae transform into pupa and then to adults. The adults swarm over the bagged material stored therein. Adults are of moderate longevity only, persisting for a period of weeks rather than months and donot feed on stored products. Only one generation is completed per year.

Spider beetle : *Gibbium psylloides* (de Czenpinski)

Coleoptera : Ptinidae

Distribution : This pest originated in Australia. It is a cosmopolitan species which reached Europe in the 19th century and is now widely distributed.

Identification : Mite-like appearance, 1.7-3.2 mm long; colour, shiny red-brown to black; lacks characteristic waist like constriction; devoid of hair.

Host : Spider beetle larvae will infest all dry animal and vegetable matter including grain, spices, fish meal, dog biscuits, dried fruits and a wide variety of miscellaneous debris.

Nature of damage : They scavenge among the debris of neglected stores and have even been reported infesting insecticidal derris powder. The larvae bore holes in which they pupate, and in so doing may damage packaging or the commodities themselves. They also contaminate commodities with droppings and silk webbings. Adult beetles too will damage packaging including bags and sacks. Gregarious and nocturnal, they spend the day in cracks and crevices amongst packaging and the fabric of the store. Consequently, they can thrive in old buildings where they find many harbourages.

Habits and Life History : The female spider beetle may produce 120 or more eggs over 3-4 weeks in early summer. They are laid either singly or in batches and being sticky, adhere to the substrate. At 20-25°C the eggs hatch in 3-16 days, giving larvae, which are fleshy, curved, covered with fine hairs and relatively immobile, although they rapidly curl up when disturbed.

Development takes at least 6 weeks during which the larvae moult 4 or 5 times. When mature, they wander in search of a pupation site where they spin a cocoon cell to pupate. These cocoons are often found on the outside of cartons through which the larvae have bored. Adults emerge after 20-30 days and will live for as long as 12 months. If disturbed they will feign death. The full life cycle takes 3-6 months depending upon ambient conditions and can take place in a temperature range of 5-28°C.

Saw toothed grain beetle : *Oryzaephilus surinamensis* (Linn.)

Coleoptera : Silvanidae

Distribution : This pest was described from specimens from Surinam. It is cosmopolitan in distribution and found all over India.

Hosts : Grain and grain products, nuts, processed food, flour, maida, biscuits, dry fruits, oil seeds, oil seed cakes, spices, cocoa beans etc.

Identification : Adult is dark brown, much flattened, slender beetle, about 2.5-3.5 mm long; neck shield with two deep longitudinal grooves and six saw-tooth like projections on each side of thorax. The posterior femora of the males are armed with a tooth, while those of the female are not. Larva is pale cream with two slightly darker patches on each segment, freely mobile and grows to a length of 3.5-4.0 mm.

Nature of damage : Adults and larvae cause roughening of grain surface and off odour in grain. Grains with higher percentage of broken dockage and foreign matter sustain heavy infestation which leads to heating. They prefer the germ portion of the seeds.

Habits and life history : Female lays about 300 whitish slender eggs loosely in cracks of storage receptacle or godown. Eggs hatch in 3-5 days in summer but in cooler weather they may not hatch for about a fortnight even. The larval period lasts from 14 days to as many as 7 or even 10 weeks depending on the temperature and other weather conditions. Full grown larva makes protective cocoon like covering with sticky secretions. Pre-pupal and pupal periods last for 7-21 days. Under most favourable conditions, the development of the eggs up to stage of emergence of the beetle may be completed in 22 days. Maximum longevity of the adults is reported to be 3 years. Generally, they survive for 6-10 months. Multiplication is quick in rainy season and in coastal areas.

Parasites and predators : The parasitic wasp, *Cephalonomia tarsalis* Ashmead and various protozoan pathogens, *Mattesia oryzaephili* and *Nosema whitei* are known to attack this species.

Yellow meal worm or European meal worm : *Tenebrio molitor* (Linn.)

Coleoptera : Tenebrionidae

Distribution : Cosmopolitan

Hosts : The pest is known primarily to feed on flour and meal but also feed refuse grain, coarse cereal products and other milled products that accumulate under the bags, bins etc. because the adults show preference for damp and moulded materials which can be easily found in accumulations at such places.

Identification : The adult is black, about 12-17 mm in length with well developed wings. Males and females superficially look alike but can be differentiated by the structure of genitalia. The larvae are pale yellow in colour.

Nature of damage **:** This pest prefer to feed on dacaying grain or milled cereals in damp and poor conditions. This insect is usually found in places not frequently disturbed such as dark corners, under sacks, in bins and feed in stores.

Habits and life history : The insect generally lives in dark and dingy places in godowns, neglected corners in mills and under bags containing sweepings. Their presence would always indicate bad hygiene of the godown. A single female on an average deposit about 75 eggs which are laid loosely in small clusters and continues laying eggs up to a period of three weeks or so. Sometimes, more than two months may be the frequency of oviposition. When first laid, the eggs are covered over with a sticky secretion that causes them to become quickly coated with the particles of food in which they are deposited. The eggs are oval, opaque and whitish, measuring 1.5 mm in length and 0.6 mm in width. The incubation period varies from 7-10 days depending on temperature. The newly hatched larvae are white which turn to yellowish brown with age.

Parasites and predators : *Tenebrio* spp. are known to act as intermediate hosts to the rodent tapeworm, *Hymenolapis diminuta.* Many species of pathogenic protozoa have also been isolated from meal worm populations and ichneumonids such as *Pimpla* spp. parasitize larvae of this pest.

Rust red flour beetle : *Tribolium castaneum* (Herbst)

Confused flour beele, *T. confusum* Jacq.

Coleoptera : Tenebrionidae

Distribution : Cosmopolitan

Identification : The adult is a small reddish brown beetle and slightly flattened dorsoventrally. It measures about 3.5 mm in length and 1.2 mm in width. Its antennae are bent and bear a distinct club formed by the three enlarged terminal joints. The last antennal segment transversely rounded. The newly hatched larva is brownish white in colour, cylindrical and wiry in appearance. The body segments have a large number of fine hair and the terminal segment has a pair of spine like appendages. The full-grown larva is pale yellowish in colour and measures about 5 mm in length.

An allied species, confused flour beetle, *Trobolium confusum* is slightly larger than *T. castaneum*. The compound eyes are completely notched antennae gradually enlarge toward tip. This insect is a poor flier and found in imported grains or hilly regions.

Host : Broken grains, processed food and milled products, like flour, *maida, suji etc.*

Nature of damage : The damage is caused by the grubs as well as adults. They generally do not feed on the sound grains but are serious pests of milled products. The larvae are negatively phototactic and are always found hidden in the food. A heavy infestation of product by these insects gives a yellow colour to the food which becomes mouldy with the excreta of larvae and adults and has a pungent, disagreeable odour making it unfit for human consumtion.

Habits and life history : The rust red flour beetle is very active during April to October and hibernates in the adult stage during winter. A single female lays from 300 to 450 eggs in grain mass, flour or other food stuffs. The eggs are white, transparent, cylindrical and rounded at both ends. The hatching period lasts 5-12 days. The worm like larvae undergo 6-7 moultings and become full grown in 22-25 days at 30°C temperature. Pupation takes place in food stuff without formation of cocoon. The pupal period lasts 6-9 days. The developmental period (egg to adult stage) is 26-30 days in summer but is longer under unfavourable conditions of temperature, humidity and food. There are 5 to 6 generations in a year.

Parasites and predators : Pyemotes ventricosus, Acrophenax tribolii and other predatory mites feed on eggs and young larvae.

Long-headed flour beetle: *Latheticus oryzae* (Waterhouse)

Coleoptera : Tenebrionidae

Distribution : It is essentially a tropical species originating in India but now cosmopolitan as a minor secondary pest of cereal grain products.

Hosts : Flour of wheat, rice, corn, barley, cowpea, *maida*, pearl millet, rye and green gram.

Identification : It is yellowish brown beetle with an elongate head. Antennae nearly as long as head, the two basal segments not visible from above. The third segment narrowest, about as long as broad; fourth, fifth and sixth segments transverse, each slightly broader than the proceeding, the seventh distinctly larger than sixth, eighth largest, ninth and tenth slightly narrow than eighth, eleventh segment narrowest, somewhat flattened. The larvae are small white vermiform and cylindrical with slightly scleretized head capsule.

Nature of damage : The infestation of this pest is usually of a secondary nature, generally found in association with *Tribolium* sp., *Oryzaephilus* sp. etc. Adults and larvae when present in excessive numbers may also induce heating. It is reported to be particularly to breed in high carbon dioxide concentration.

Cadelle beetle : *Tenebroides mauritanicus* (Linn.)

Coleoptera : Trogostidae

Distribution : It is cosmopolitan in distribution.

Hosts : Rice, wheat, maize, groundnut, copra, oilseed cakes, cocoa beans, dried fruits, various seeds, spices and dried vegetables.

Identification : The beetle is elongate oblong, flattened and shining black or reddish brown in colour. It is about 1.27 cm in length and is said to be the largest in the stored grain insect pests of the world. Antenna swollen abruptly, 8 to 10 mm long, last 4-5 segments of antenna swollen. The larva is creamy white with black head.

Nature of damage : Although, it is basically predacious, usually feeds directly on cereals and seeds both as adults and larvae. Farmers of the wheat belt in USA call the larvae 'flour worm'. They store sound wheat grains and after some time much of which is reduced to flour dust. It is a germ feeder but eats on the endosperm too. In wooden bins, this insect makes tunnels and cut holes and hides there till new grain is filled us. It feeds on rice weevil but cannot be used as predator as it feeds on grain even in the presence of rice weevil. The larvae and adults gnaw holes through bags, cardboard, paper containers, bamboo baskets etc. Both larvae and beetles are commonly found in the accumulation of stocks in the bottom of storage containers, particularly in wooden stores and bamboo baskets in valley of Kumaon hills (India).

Habits and life history : The adults and larvae are predatory on other stored product insects and both stages can be long lived, larvae appearing to have a diapause like arrest permitting them to overwinter and adults being able to survive upto 22 months. During this long life cycle, females may lay over 1000 eggs over a one year period, the pre-oviposition period varying between about 2 weeks in summer to upto 6 months in adults emerging in autumn, suggesting a reproductive diapause in addition to the possible diapauses that larvae may enter in autumn. Larvae may damage structural materials such as soft wood or plastics in boring out a chamber for pupation. In the absence of developmental arrests, development may be completed in as little as 10 weeks towards the upper end of the 20-30^0 C developmental range. Under unfavourable conditions, it may take 3-4 years to become full grown. Most larvae moult 3 to 4 times but under unfavourable conditions, they may moult several times. Larva pupates in the tunnel or hole made inside the grain or other parts of the store. Pupa is white and can be seen in between the two woven seeds. Pupal period lasts for about 10 to 12 days. Cadelle larva and adult both can live hidden deep in the holes for long in cold winters. They are true hibernating insects.

Parasites and Predators : Pyemotes ventricosus and other predatory mites attack eggs. Larvae are attacked by the bethylid, *Cephalonomia* spp. Two adult parasites, *viz., Nosema tenebrionidae* and *Hymenolepsis diminuta* have been reported.

Potato tuber moth : *Phthorimaea operculella* (Zell.)

Lepidopera : Gelechiidae

Distribution : This pest occurs especially in hot and dry climates.

Hosts : It is destructive to stored potato tubers and also attacks tobac co, tomato, brinjal and solanaceous weeds.

Identification : The adults are very small narrow winged, night active moths, greyish brown in colour and mottled with darker brown. The forewings are light black in colour with fringed apical margin. A characteristic feature of the moth is that its hind wings are densely fringed. Female moths are larger (8.2 mm) than male moths (7.63 mm). Moreover, newly emerged females has a dark black patch of scales in the middle of forewings. The abdominal tip is conical in females and broad in males.

Nature of damage : The infestation starts in the field on leaves and acts as an initial source of infestation. Potato tuber moth is a serious pest of stored potato tubers and does considerable damage to potato plants and developing tubers in field. The damaging stage of the pest is larva, which feeds on potato foliage and attack tubers in the field before and shortly after harvesting. The foliage feeding larvae make transparent galleries in infested leaves. The tuber mining larvae bore mostly near the eyes of tubers filling the tunnels with excrement. Fungus grows in burrow and discolours it so that course of work may be easily followed. Latter, the skin of potato partially dries and sinks so the scars become very prominent; this is called sub-epidermal injury.

Habits and life history : Moths emerge from overwintering larvae in early spring and lay eggs chiefly on underside of leaves or upon exposed tubers. The eggs hatch in 4 – 7 days. Newly hatched larva is white with dark brown head. Full grown larvae are about 12 mm in length and 3 mm in thickness. It is pink in colour with dark colouration on both sides. The larvae are full fed in 15–20 days. They pupate in a greyish, silken, dirt covered cocoons entangled in dead leaves or among trash etc. on ground and emerge as adults in 6-8 days. An entire generation may develop in 22–25 days in warm weather and 5–6 generations are completed in a year in Kangra valley of Himachal Pradesh. The damage is most severe in years of low rainfall and high temperatures. In Assam, as many as 11 generations, in the Northern plains 8-9 generations and in plateau region 10-13 generations are reported to be completed in a year. In Bangladesh, 18 generations are completed in a year.

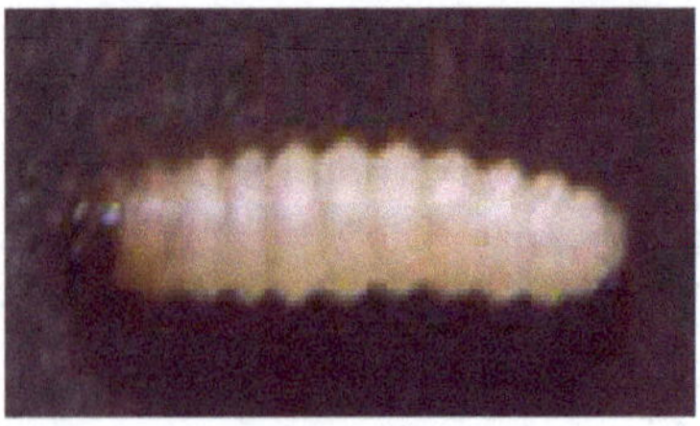

Fig : Larva of *Phthorimaea operculella (Zell.)*

Fig : Moth of *Phthorimaea operculella (Zell.)*

Pink bollworm : *Pectinophora gossypiella* (Saunders)

Lepidoptera : Gelechiidae

Distribution : It is found in America, Africa, Australia and Asia

Hosts : Cotton seeds.

Identification : The adult is deep brown moth measuring 8-9 mm across the spread wings. There are blackish spots on the fore wings and the margins of the hind wings are deeply fringed. The full grown larvae are pink (8-10 mm long) which are found inside the seeds. The infested seeds lose viability.

Nature of damage : The damage is caused by the caterpillars only. They are internal feeders of seeds. The source of infestation is the fields where they cause damage to the flower buds, panicles and bolls of cotton or the fruits of okra and other allied plants.

Habits and life history **:** The yearly life cycle begins with the emergence of moths in the summer. The emergence takes place at two distinct times; in May-June and then in July-August. The female lays whitish flat eggs singly on the under side of the young leaves, new shoots, flower buds and young green bolls of cotton in the field. The eggs hatch in one week and caterpillars on emergence are white and turn pink as they grow older. Soon after emergence, the larva enters the flower buds, flowers or the bolls. The holes of entry close down but the larva continues to feed in the seed kernels and reaches in the storage when seeds are stored. The full grown caterpillars come out for pupation in the ground, fallen leaves and debris. By October-November, 4-6 generations are completed in the fields. Full grown larvae of the last generation do not pupate and hibernate in the winter. Few of them are picked along with the healthy cotton and reach to the ginning factories from where just a few return to the fields along with the seeds. The hibernating larvae lie curled inside the double seeds (two seeds join together as a result of infestation) for many months after passing the winter, they emerge as moths. The last life cycle is very long, covering 5-10 months; although, during the active season, the life cycle is short, taking only 3-4 weeks.

Parasites and predators : The egg parasitoid associated with the pest is *Trichogramma achaeae* Nagaraja and Nagarkatti and the larval parasitoids include *Apanteles angaleti* Muesebeck and *Bracon greeni* Ashmead.

Angoumois grain moth : *Sitotroga cerealella* (Olivier)

Lepidoptera : Gelechiidae

Distribution : The moth takes its name 'Angoumois' after the province of Angoumois in France from where it was reported in 1736 as a pest of wheat. The distribution of this pest is worldwide. In the Indian subcontinent, it is more abundant in the mountainous areas or where the climate is rather mild.

Hosts : Paddy, jowar, maize, barley and wheat grains.

Identification : The moth is buff coloured with a wing span of 13-19 mm and a length of 6-9 mm, pointed hind wings bearing a heavy fringe of hair, most prominent along the posterior margin. Labial palps sickle shaped, curving upward with last segment needle like and commonly black in colour. A full grown larva is 5 mm long with a white body and yellow brown head.

Nature of damage : The insect is considered to be the second important pest after rice weevil. The damage is caused by the caterpillar stage only and the damage is at maximum during the monsoon. Only the larva is capable of tunneling the sound grains. As a result of attack, the grains are hollowed out by the larvae and filled up with the excreta and webbing. The damage restricts to top or outer layers of the grain only. Generally, one caterpillar is found in one kernel. It is difficult to spot out the kernels having larvae inside because the larvae after entering the grain close the entry by spinning a silken web. Faecal contamination, silk produced by young larvae and scales from the wings and the body of the adults led to loss of grain quality. Heavily infested grain possesses a sickening smell and taste thereby making it unfit for human consumption. The insect also attacks the ripening seeds in the field.

Habits and life history : The insect commonly infests produce before harvest and is abundant only in surface layer of bulk stored grains in the store. The pest is active from April to October. The insects hibernate in larval stage in the winter, as the season warms up pupate in early spring and start emerging as adult in the March. After emergence, moths mate within 24 hours and the female's start laying eggs singly or in small clusters in cracks and crevices of the floor or attached to the grains in the store or on ear heads in the field. Each female is capable of laying nearly 300 eggs. The eggs are small and white when freshly laid, turning reddish later on. The incubation period lasts for 4 to 10 days. On hatching, the worm like tiny caterpillars bore into the grains and feeds on its contents and remains there for rest of their life. The larvae become full fed in 20- 30 days after moulting five times. Full fed larvae measure 5-6 mm in length and have a white body with yellowish head. Before pupation, the larvae construct a silken cocoon in a cavity made during feeding and then turn into reddish brown pupa. The pupal stage lasts for 7 to 10 days. The adults come out through a round emergence hole.

Parasites and predators : Parasitic wasps such as *Pteromalus pyrophilus* Kollar and *P. cerealellae* (Ashmead) attack larvae and *Trichogramma cacoeciae* Marschal attack eggs. Mites, like *Pyemotes ventricosus* also attack eggs.

Fig or Almond moth : *Ephestia cautella* (Walker)

Lepidoptera : Pyralidae

Distribution : This pest is cosmopolitan in distribution.

Hosts : Dried fruits such as currants, raisins, dried apple, dates, berries, figs, almonds, walnuts, tamarind seeds etc. It has also been recorded on lac, malted milk, dried mango, garlic bulbs, groundnuts, cocoa, beans, various cereal grains and grain products.

Identification : The adult is a greyish coloured moth and measures about 12 mm across the spread wings with transverse stripes on grey coloured wings. In the resting position the fore part is elevated giving a distinct slop to the wings, which are wrapped about the body. The adult is slightly smaller than *Corcyra* but longer than *Sitotroga*. The larva is white in colour with pinkish tinge and possesses setae arising from pigmented spots on the cuticle.

Nature of damage : The damage is caused by the caterpillars, which infest the germ portion of sound kernels. One caterpillar in its life time destroys the germs of about 48 kernels of wheat. The larvae spin silken threads profusely wherever they crawl. Food particles, grains, flour etc., all are webbed together in a connected mass. Sometimes, the machinery of flour mills is clogged as a result of cake formation of flour by larval threads. Imported varieties of wheat are more susceptible than the indigenous ones to their attack. The greatest damage is caused during the rainy season.

Habits and life history : The pest is active throughout the year but their population is at peak during the rainy season (July to September) and minimum during the winter months (November to February). The adult avoids strong light, rests in dark places during the day time and has a flight rhythm with a period of active flight during 5 to 7 pm and 6 am, when daily fluctuations in temperature and humidity occur. A single female lays from 200 to 250 eggs which are small and whitish, indiscriminately scattered in stores, in cracks, on grains or other surfaces. Eggs hatch within 3 to 6 days and measures less than 1 mm in size. The larvae start feeding on grain debris. The newly hatched caterpillar starts spinning silk profusely and makes small silken tubes among the food particles or grains in which it remains lodged and grows. The larval stage lasts for 30 to 50 days. The full grown larva measures 6 to 12 mm. long and is white in colour with a pinkish tinge. The full grown larva spins a cocoon and pupates therein. Pupal life is therein 7 to 10 days. The developmental period of one generation lasts for 6 to 10 weeks. There are 5 to 6 generations in a year.

Parasites and predators : The parasitic wasps, *Trichogramma pretiosum* (Riley) and *T. evanescens* Westwood attack eggs; the braconids, *Hebrobracon brevicornis* (Wesmael) and *H. hebetor*

(Say) and the ichneumonid, *Venturia canescens* (Gravenhorst) attack larvae.

Indian meal moth : *Plodia interpunctella* (Hubner)

Lepidoptera : Pyralidae

Distribution : This species, first recorded on Indian meal, is a pest of a very wide verities of stored products. Now, it is distributed worldwide.

Identification : The moth is about 12-13 mm in wing expanse. Basal half of fore wing is silver white or greyish, outer 2/3 portion is reddish copper bronze lustre with irregular bands. Hind wings long silvery grey with silky fringes. The thorax is slightly darker with reddish scale. The palps form a characteristics cone like beak in front of the bead.

Hosts : Dry fruits, cereals, groundnut, cereal products, meals, soybean, nuts, cocoa, beans, chocolate, herbs, dead insects etc.

Nature of damage : The larvae feed first on the embryo of the grains and while eating, spin a silken thread on which the larval droppings as well as particles of the produce accumulate. The adults fly from one bin to another and spread the infestation.

Habits and life history : It breeds throughout the year under favourable warm conditions; however, in cold weather, the insect over winters in larval stage. Upto 300 eggs are laid by the female but the precise number depends on the food taken by the female during larval life. The eggs are minute, white, oval shaped, laid singly or in clusters on or near the appropriate foodstuffs. The incubation period is 2 days to 2 weeks depending upon weather. The larvae feed upon the grains or other foods and become full grown in 30-35 days. They crawl upto the surface of the food material and pupate within a thin silken cocoon. The pupal stage lasts 4-35 days. In summer, the life cycle is completed in 5-6 weeks and there are about 4-6 generations in a year.

Parasites and *predators : Tricogramma pretiosum* (Riley) and *T. evanescens* are egg parasites while *Hebrobracon hebetor* and *Venturia canescens* attack larvae. Larvae are susceptible to polyhedrosis and granulosis viruses and to *Bacillus thuringiensis.*

Rice moth : *Corcyra cephalonica* (Stainton)

Lepidoptera : Galleriidae

Distribution : The rice moth is widely distributed and most common in tropical regions such as Asia, Africa, North America, Europe, Pakistan and India.

Hosts : Rice, sorghum, maize, wheat, groundnut, cotton seed, cocoa, palm kernels, copra, dried fruits etc.

Identification : The adult is a pale, greyish brown moth with a wing span of 14-24 mm and little larger than *S. cerealella*. The forewings appear broad with distinct shoulders. The moth may also be recognized by its labial palps which point straight forward and are blunt and inconspicuous in males and long and pointed in females.

Nature of damage : Only the larva is responsible for damage. Young larva feeds on the broken grains or on the grains which are already damaged by other insects. As the larva grows old, it becomes capable of feeding on entire grain. When infestation is high, entire stock of grains may be converted into webbed mass. Ultimately, a characteristic foul odour develops and the grains are rendered unfit for human consumption.

Habits and life history : The pest is active throughout the year but its population is at peak during the rainy season and minimum during the winter months. The insect over winters in larval stage and as the season warms up it pupates in early spring. The moths emerge in the month of March. The sex pheromones have been released by both the sexes. The male pheromone has been identified as E, E-3, 7, 11-trimethyl-2, 6, 10-dodecatriental (farnesal) from wing glands. This attracts females from short range and females also release a short range attractant, 6, 10, 14-trimethyl-2-pentadecanol after locating a male. They are active at night and a single female lays 90-200 eggs during its life span of 2-4 days indiscriminately anywhere on bags, walls, loose grains, floor etc. Eggs are small, oval and elliptical and measures 0.5 x 0.3 mm. The eggs hatch in 4-7 days and the larvae feed under silken web like shelters, preferring the partially damaged grains. Larval life is 4-5 weeks, after that it pupates inside the silken cocoon for 8-10 days and moth emerges. The developmental period of one generation occupies 4-7 weeks. There are 4-5 generations in a year.

Parsites and predators : Many species of parasitic wasps have been associated with *C.cephalonica. Trichogramma* spp. attacks eggs while *Antrocephalus aethiopicus* Masi and *A. mahensis* Masi and the braconids, *Habrobracon brevicornis* (Wesmael) and *H. hebetor* (Say) attack larvae. Larvae are also susceptible to strains of *Bacillus thuringiensis*.

Greater wax moth : *Galleria mellonella* (Linn.)

Lepidoptera : Galleriidae

Distribution : It enjoys world wide distribution and its ravages have been noted from the very early days of Aristotle (384-322 B.C.). In South India, it causes severe damage in the plains and lower altitudes but it is rare in high altitudes.

Host : Combs of all the species of *Apis* are freely attacked.

Identification : Adults are brownish grey, 10-18 mm in length and 25-40 mm across the spread wings, the female larger than the male. The colour and size of adults vary a great deal in accordance with the food eaten during the larval period; dark brown combs containing pollen give rise to darker and bigger adults. The outer margin of front wings of the males has a semi-lunar notch, whereas, the females is smooth. Freshly hatched larva is white and 3 mm in length, full grown is dirty grey and upto 30 mm long; its brownish head is conspicuous. They live in the silken tunnels constructed by them.

Nature of damage : The caterpillars eat away the combs or damage them by making silken galleries with their side tunnels near their midribs. The first indication of the entry of larva in the comb is the presence of small masses of minute particles of wax out side the holes. Latter, faint webbings are perceptible over some cells of the comb. When infestation is prolonged far enough, silken tunnels with caterpillars wriggling in them are noticed and eventually the whole comb is a mass of webbings in which the excreta of caterpillars is enmeshed. In severe cases of infestation, further brood rearing is stopped, field work is virtually suspended and the colony deserts its home to seek its fortune elsewhere, leaving it to the intruders.

Habits and life history : The pest is active from March to October. The emergence of adults starts in March and April. The adult female moth enters the hive during night or in case of weak colonies even during day time and lays eggs in the cracks and crevices of the hive and comb. About 200-800 eggs are laid in batches by a single female in the course of a fortnight. The eggs are creamy white in colour and hatch in 8 to 10 days into tiny white caterpillars about 3 mm long. Larval period extends over a month. The pace at which caterpillars grow, even in warm localities, considerably slows down in winter In stored combs, the pest hibernates in the caterpillar stage (about 70 per cent) and in the pupal stage (about 3o per cent). No adults are found in the stored combs when the pest is under hibernation. Pupation takes place in a white silken cocoon among the debris of the combs in the corners and crevices of the hive. Pupal period is about 8 days. The adult moth is brownish grey in colour, 10 to 15 mm long and the outer margins of the fore wings of the males have a semicircular notch, whereas, they are smooth in females. Mating takes place soon after emergence and eggs are laid in a day or two after mating. There are several overlapping generations of the pest in a year depending upon temperature, the availability and nature of food and whether the pest is in an inhabited hive or stored combs. In localities with comparatively warm winters, all stages of the pest (adults also) are met within hives throughout the year.

Parsites and predators : The caterpillars are parasitized by a Braconid, *Apanteles galleriae*. The female parasites lay their eggs inside caterpillars and the grubs feeds inside them

Webbing cloth moth : *Tineola bisselliella* (Hummel)

Lepidoptera : Tineidae

Distribution : Cosmopolitan

Host : Clothing, carpets, hair, fur and other furnishing household.

Identification : Adult webbing cloth moths have a wing spread of about 1.5 cm and that of the male is some what less. The body is about 0.75 cm long with wing folded and golden yellow with a satiny sheen. A tuft of reddish golden hairs on the head is upright. Larvae are shiny, creamy white with a brown head.

Nature of damage : The larva feeds on all keratin containing materials, like woolen clothing, upholstery and carpets as well as other animal based material, such as hair or fur. It lives in a silken tube, and when gregarious this can develop into dense matting.

Habits and life history : Cloth moth rarely flies to lights at night and instead prefers darkness, such as a closet or storage chest. The adults are very active, can penetrate through surprisingly narrow cracks and can fly considerable distances. Gravid females are weak fliers. Any cloth moths fluttering around the houses are probably males, because females travel by running, hopping or trying to hide in the folds of clothing. The female dies after laying about 40 to 50 eggs, singly or in groups of 2 or more, to the threads of infested clothes over a period of 2 to 3 weeks. The eggs hatch within an average of 4 to 10 days in summer but takes as long as 3 weeks in winter. The number of larval moults can vary from 5 to 45 and the period required for larval development can vary from 35 days to 2.5 years depending on the availability of food as well as relative humidity and temperature. The full grown larva is shiny, creamy white and about 12 mm long. When preparing to pupate, it spins a pupal case of silk about 8 mm long, again incorporating textile fibres and excrement. The pupal period is 8-10 days in summer and 3-4 weeks in winter. The length of the life cycle varies from 50 to 90 days, but can be extended to as long as 4 years under unfavourable conditions. There are about 2 generations a year.

Psocids (Grain lice) : *Liposcelis divinitorius* (Muller)

L. transvallensis (Endernein)

Psocoptera : Psocidae

Distribution : They are found in a variety of natural terrestrial ecosystems.

Hosts : All starchy materials

Identification : They are tiny, pin-head sized and pale grey or yellowish white in colour. They are soft bodied louse like insects with fairly large head, poorly developed eyes and slender filiform antennae.

Nature of damage : Infestation of this pest is only indicative of unhygienic, damp warehouses, where aeration is ignored and is usually associated with high infestation or mouldy conditions. Infestation of psocids persists even after fumigation when other pests are controlled, thus causing irritation or nuissance besides contamination. These are scavengers affecting only germ portion in heavy infestation. Normally they thrive on insect fragments and broken grains.

Habits and life history : About 7-60 eggs are laid which hatch to nymphs. In summer, the developmental period from egg to adult is about 20 days. The newly hatched nymphs resemble the mature insects but are smaller and lighter in colour. Heavy populations can develop in stored wheat, particularly, if it has 12.5 per cent of moisture content and the ambient relative humidity is 65 per cent.

Cigarette beetle, Drug store beetle, Coffee bean weevil, Lesser grain borer, Pulse beetle, Groundnut beetle, Rice weevil, Rusty grain beetle, Flat grain beetle, Khapra beetle, Black carpet beetle, Spider beetle

Fig. 1: Insect pests of stored products.

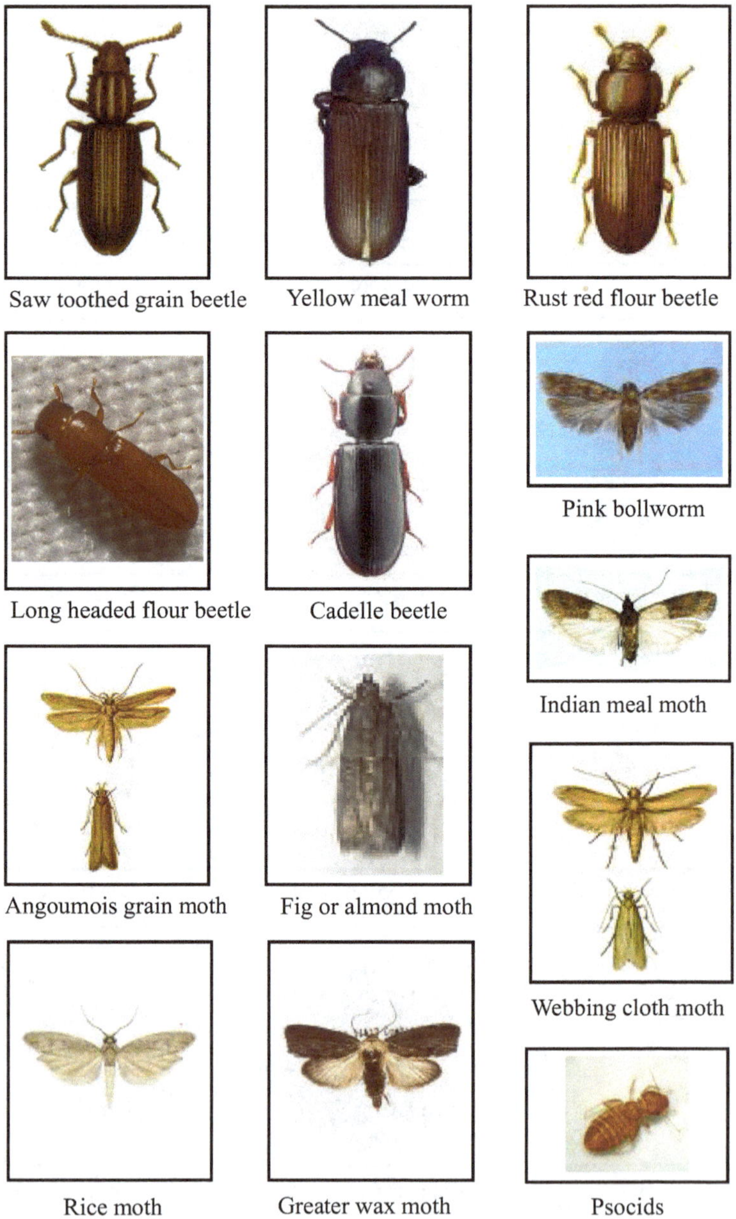

Fig. 2: Insect pests of stored products.

Chapter 5

Mite Pests of Stored Food Products

Acarid Mites (Acaridae)

- Grain mite, *Acarus siro* L. (= *Tyroglyphus farinae* De Geer)
- Cheese mite, *Tyrophagus casei* (Oudeman)
- Mould mite, *Tyrophagus putrescentiae* (Schrank) (= *lintneri* Osborn)
- Dried fruit mite, *Carpoglyphus lactis* (L.) (= *Acarus passularum* Hering)
- Grocer's Itch mite, *Glycyphagus domesticus* (De Geer)

There are several species in this cosmopolitan family that are of importance chiefly as pests of stored foods, flour, grain, seeds and bulbs, feeding on these products or the fungi that develop on them. These mites are mostly very minute, almost colourless, with the body bearing a few long setae, and with the legs terminating in one claw and usually a sucker. When very abundant the mites and their cast skins form a buff-colored dust and give rise to a characteristic odour. The living mites move slowly away from a source of light. An interesting characteristic of these mites is a nymphal resting stage (hypopus) during which they take no food but attach themselves to other mites, insects and humans by means of suckers and are thereby transported to other suitable places for development, where they drop off, moult, feed, and develop into adult stage. The hypopus may be resistant to fumigants. However, the cheese mites, *Tyrophagus casei* do not have a hypopial stage. Prolonged storage of products subject to infestation without cleaning of pantries and shelves where such materials are placed. These mites can subsist on organic debris in cracks, crevices and corners and are readily spread to other areas.

Grain Mite: *Acarus siro* (L.) (= *Tyroglyphus farinae* De Geer)

The grain mite is the most important and best known of the mites infesting grain and flour but can also infest other stored food products. In severe infestation, grain mites are responsible for the mite dust on the surface of cheese or on the floor at the bases of stacks of flour or wheat sacks. Although the bodies of the mites are almost colourless, their legs are brown, so the mite dust has a brownish tinge. If the mites are crushed, their minty odour is easily recognized. *Acarus siro* is widely distributed in temperate zones, it occurs in the tropics only through repeated introductions with grains and foodstuffs. It is not an important pest in hot climates, if it persists at all.

Identification and life cycle : This mite is pale straw to dark reddish brown, with 4 pairs of legs varying in colour from pale yellow to reddish brown. The male is 0.32 to 0.42 mm long and the female is 0.35 to 0.65 mm. The mite can be distinguished from members of other genera because of the femur of the male's foreleg being enlarged and bearing a stout spine ventrally, a characteristic of the genus *Acarus*. Pile formation on floor is typical of mite infestation and in bagged storage dust formation is characteristic. If an infested sample is kept in hand, itching sensation is the sign of = presence of mites.

In a day, a female may lay 20 to 30 oval, smooth, white eggs, 0.12 mm long, singly or at random on the food material. Under favourable conditions, the female can lay up to 800 eggs in her life time (about 40 days). The larvae feed and grow rapidly for 3 days, then lie motionless for 2 days before molting. The life cycle may require 17 days at a favourable temperature ranging from 18 to 22°C but during the winter at 10 to 16°C, it requires 28 days. *Acarus siro* does not have tracheae, so the cuticle must be permeable to gases and therefore also to water vapour. Although the epidermis and cuticle obviously have some ability to regulate water balance, this does not suffice to prevent a lethal rate of water loss below 60% relative humidity, and the mites cannot survive at such humidities. The distribution of the grain mite in stored grain is conditioned by the relative humidity of the air between the kernels, which in turn depends on the water content of the grain. The mite avoids relative humidities below 75 per cent and above 85 per cent.

The grain mite is commonly found in bird nests around houses and stores, and it is believed that birds aid in its dispersal. Rodents and other mammals also probably contribute to dispersal, as well as human on their shoes and clothing. The mites are also found in grocery stores, and are usually the culprits responsible for a severe dermatitis known as 'grocer's itch'. In fact, probably any acarid mites could cause dermatitis on the body of a sensitive person.

Cheese mite : *Tyrophagus casei* (Oudeman)

These cosmopolitan mites are common in stored food, cheese, grain, damp flour, old honey combs, and insect collections. They can eat out small holes in cheese, and have been cultured for addition to Altenburger cheese to impart a characteristic piquant taste. When the cheese is covered with a greyish powder, consisting of enormous numbers of living and dead mites, cast skins, and faeces.

Identification and life cycle : This species, like others of the genus, has at least 4 pairs of long bristles trailing from the hind end of the body. It is translucent, but has tan-coloured legs and mouth parts. The males are 0.45 to 0.55 mm long and the females 0.50 to 0.70 mm long. The life cycle requires 15 to 18 days at 23°C (73 °F) and 87 per cent relative humidity.

Mould mite : *Tyrophagus putrescentiae* (Schrank) (= *lintneri* Osborn)

This is one of the most destructive pests of stored products. The mould mite is found most frequently on products having a relatively high fat content, such as linseed, dried egg yolk, peanuts, cheese, ham, herring meal and copra but also occurs on dried banana, spilled wheat, oats, barley, and flour, and is also a pest on mushrooms. The mites sometimes cause considerable concern and annoyance when infesting stored products in homes and responsible for the severe dermatitis called grocer's itch when attacking human who handled infested materials. The mites also cause copra itch familiar to persons handling infested copra dust. As already stated under *Acarus siro*, itches can also be caused by other acarid mites, among which *A. siro* is very notable.

Identification and life cycle : This is a small, translucent mite, with almost colourless chelicerae and limbs. The males are 0.28 to 0.35 mm long and the females 0.32 to 0.41 mm. It is more slender than other acarids and the long setae do not project so stiffly from the surface of the body. The life cycle requires 2-3 weeks at 23°C and 87 per cent relative humidity.

Dried fruit mite : *Carpoglyphus lactis* (L.) (= *Acarus passularum* Hering)

This species is found principally on dried fruits, jams, and products containing lactic and acetic acids. It has been found in honey combs, fruit drink residue, senna confections, wine clinging to pieces of cork or forming a floating scum on the surface, rotting potatoes, flour, dried milk powder. It seems to be particularly attracted to substances that start to ferment. This species is sometimes found in enormous numbers on a variety of dried fruits, imparting a disagreeable odour. It has also been observed on cured hams and the pollen paste of bees and can cause dried fruit mite dermatitis.

Identification and life cycle : This oval, slightly flattened mite is translucent, except for the colour of the contents of the alimentary canal, which can be seen through the body wall. The legs and mouthparts are pinkish. There is no transverse suture dividing two body regions as with certain other species. The two sexes are 0.38 to 0.40 mm long, and are similar in appearance. The females can lay 25 - 30 eggs in a week attaching them to the substrate by means of a stalk and can then survive another 35-45 days.

Grocer's itch mite : *Glycyphagus domesticus* (De Geer)

This mite is also called house or furniture mite, is world-wide in distribution, being found in flour, wheat, hay, linseed, tobacco, sugar, cheese, and bee frames, in bees' and birds' nests, and in the sweepings from barns. It may increase to enormous numbers on rush furniture and on upholstered chairs stuffed with green Algerian fibre. As the common name implies, the mites often cause grocer's itch in common with some other acarids.

Identification and life cycle : Live mites of this species can be distinguished from *Glycyphagus destructor* by having a more rounded body and by a uniform, unhurried gait in contrast with the jerking progress of the latter. The long hairs of the body and legs are less coarsely pectinate than those of *G. destructor*, and in life they radiate stiffly from the surface. Like the related species, this mite has long legs with tapering segments. The males are 0.32-0.40 mm long, and the females are 0.40-0.75 mm in length.

Losses Due to Mites

(a) *Quantitative losses*: It includes proportion of grain damaged and weight loss caused by mites. It was estimated that mites generally eat the food equivalent to their body weight in a single day. *Acarus siro* was found to consume 3 per cent of weight in wheat before they died out.

(b) *Qualitative losses*: It includes loss in germination, nutritive value, organoleptic properties and health hazards caused by acarid mites.

- *Loss in germination*: In different pulses, 5 to 12 per cent germination loss was estimated due to *Tyrophagus putrescentiae* infestation. Due to *Lardoglyphus destructer* loss in germination in wheat was 38 per cent.

- *Loss in nutritive value and organoleptic properties:* Low fat acidity value, decrease in protein and starch content and increase in reducing sugar due to mite infestation. Mite feeding also reduced the iron content and vitamin B- complex in grain. Heavy infestation of mites led to tainting and minty smell in grains that become unfit for human consumption.
- *Health hazards:* The increasing occurrence of allergies in the developed countries and greater interest in food safety has drawn attention to the presence of allergens in agroproducts during storage. Mites can be important sources of air borne allergens, especially on farms. They cause occupational asthma by inhalation and anaphylactic reactions when ingested in high numbers. Allergic reactions due to mites are reported in persons handling the grains. Its severity increases many folds due to its perennial nature and presence of multiple allergens. Bronchial asthma and chronic rhinopathy were observed in patients associated with mite allergens. Conjunctivitis and perioral syndrome were also found to be related with stored mite infestation.
- *Interaction with other biotic variables:* Mites are excellent carriers of fungal spores known to be positively associated with storage fungi. Mites are also involved in the initiation and spreading of hot spots. Mixed population of mite, insect and fungi led to 15°C rise in grain temperature under storage conditions. Complete deterioration of grains occurs at this stage.

Management of Storage Mites

a. Preventive measures

Unfortunately, once large numbers of mites are noted, control becomes very difficult. Following are the few measures, which should be taken into consideration for mite pest free storage.

- Grains and seeds should be stored below 12 per cent moisture content.
- Stores should be properly cleaned and disinfected before storage.
- Avoid contamination during post harvest storage.
- All windows, ventilators, walls, roof, and floor must be repaired from time to time to make the store free of rodents and birds that often act as a source of mite infestation.
- Good quality of polythene may be used on floor, walls, and roofs to further protect the stored products from dampness.
- All waste products and floor sweepings must be burnt to prevent reinfestation by mites in them.

b. Curative measures

- *Chemical control*: Fumigants are ineffective on eggs and early stage larvae, which result in resurgence of mite population after few days. Moreover, mites need 1.5 times higher dose than recommended for insects. Spraying of lindane, malathion or chlorpyriphos -phoxim mixture, etrimphos and primiphos methyl is effective against storage mites. Bin wall treated with malathion provide a protective barrier against mite pests.
- *Use of inert dust*: Inert dust able to reduce the granivorous mite population in stored grains.
- *Use of photoperiod*: Photoperiod also exerts a powerful influence on their life processes as these generally remain in dark during post harvest in light exposures lead to decrease

in fecundity, egg viability, growth and increase in hatching and developmental period.

- *Use of botanicals*: Spice oils (black pepper, mint, turmeric, garlic etc), neem oil, vegetables oils and tulsi oil have inhibitory/repellent activity against acarid mites.
- *Biological control*: Biological control of mites is being exploited using several predatory mites, *viz., Cheyletus eruditus*, *C. malaccensis*, *Acaropsis docta*, *Androlaelaps carolis*, *Blattisocius* spp. and *Pyemotes tritici* are known to feed on a acarid mites.

Grain mite

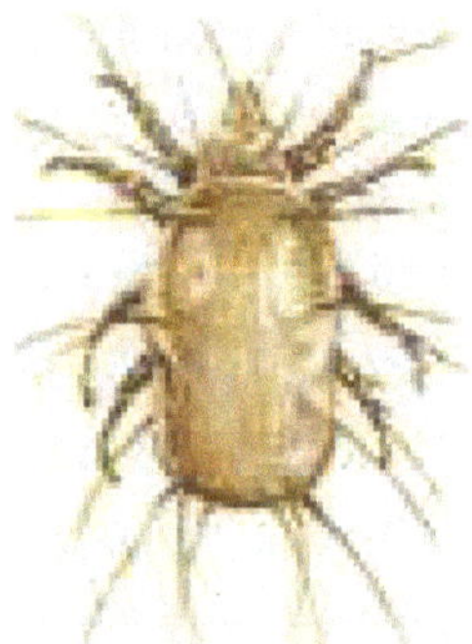

Cheese mite

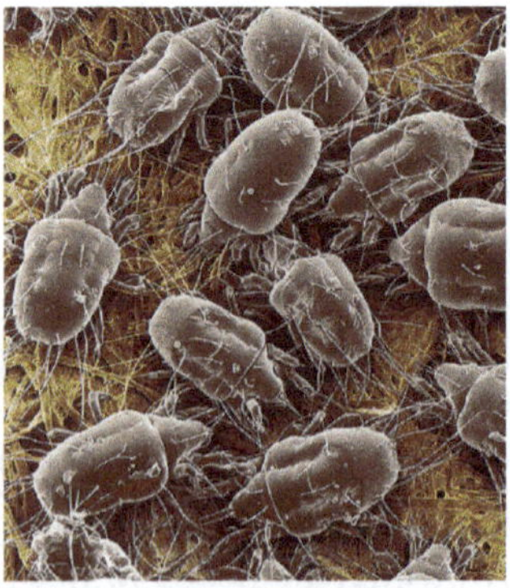

Mould mite

Grocer's mite

Fig. 3: Mite pests of stored products.

Chapter 6

Avian Pests

It is interesting to note that only 25 of about 1200 species of birds found in India have been reported to inflict damage to crops, vegetables and fruits. The exact ecological causes for population increase of individual species have yet to be studied. It would appear that, in general, the granivorous birds became abundant with the increase in area under grain crops and the subsequent storage of grain in godowns and bulk storage. The details of important bird pests are given below:

1. Crow

Classification

Order	:	Passeriformes
Family	:	Corvidae
Scientific name	:	*Corvus splendens* Vieillot
Common name	:	House crow (Hindi: *Kauwa or Kaga*)
Distribution	:	Throughout India and found up to height of about 1,200 metres.

Nature of damage : Normally a crow is omnivorous, feeding on kitchen wastes to dead animals and hence considered as the best scavenger. It feeds on the grain mixed up in refuse, sweepings and spillage around warehouses. It enters into warehouse only if left unattended for long. The maximum damage is done in the threshing yards and open storage. In fields, considerable damage is done to maize, wheat, jowar, other grains, groundnut, fruits and chillies.

Identification : It is one of the most familiar birds with grey neck and black body measuring 43 cm from beak to tail. It makes a noise *kaon, kaon, caw, caw.*

Reproduction and habits : A crow starts breeding when one year old. The breeding season is April to August extending up to January with 4-5 eggs per brood. The eggs resemble with the eggs of *koel* and are light blue to greenish with dark speckles. Nests are untidy platform of twigs, branches and sometimes wire lined with rags, fibres, grasses etc. with a central depression. At night, it rests in roosts on trees nearer to habitation. It is strong flier. It flies considerable distances for feeding in fields during day time. They live, feed and fly in large groups (especially in winter). Smaller groups are seen in summer. It is an intelligent, active, attentive, cunning and bold bird with no fear of human beings.

Management : A very wary bird; acetylene exploders have been reported effective against their depredations in orchards. Only a small percentage of a flock can be poisoned with strychnine treated grain but generally the rest of the flock will quit an area as the first starts dying. For this reason, thallium sulphate baits because of their slow action may give better kills. They can be trapped in large V-shape traps. Destroy eggs and nests during June-August. A piece of *chapatti* dipped in 0.3 per cent methyl parathion or 2 per cent fenthion placed on top of roof is also good bait.

2. Sparrow

Classification

Order	:	Passeriformes
Family	:	Ploceidae (Sub family: Passerinae)
Scientific name	:	*Passer domesticus* Linn.
Common name	:	House sparrow (Hindi: *Chidia*)

Distribution : In India, it is common on the plains and the mountains up to 2,150 metres.

Nature of damage : Most important avian pest in storage; principally granivorous; lives or enters in house and warehouses. Contents are damaged in bag storage. In open storage, maximum losses are noticed. Besides this, contamination with excreta, feathers and dead bodies occurs.

Identification

1. Sexual dimorphism is found.
2. The female is earthy brown, streaked with blackish and rufous with whitish under parts.
3. Male has a loud monotonous and aggravating song *tsi, tsi, tsi* or cheer uttered ad nauseum with fluffed out plumage.
4. Body length is 15-16 cms.

Reproduction and habits : Breeds all round the year, raising 5 broods per year. Incubation period is of 14 days. Female lays 3-5 pale white greyish eggs in one brood. They make nests out of dried vegetation, cushioned with soft feathers or tender grass leaves in holes or cracks in buildings, even on ceiling fans covers etc. Before sun set they collect in a favourate roost of a leafy tree or thorny thickets and make a lot of noise before sleep.

Management : Destroy eggs and nests during April-May and September-October. Bajra seeds soaked in 2 per cent fenthion emulsion and dried, are placed in small cups and hung from the rafters or branches of trees. Spraying the wheat crop when ears are in the milk stage with Thiurum (TMTD) 0.6 per cent, repels the sparrow and protect the crop. Repellent devices both acoustical and visual can be used at the time they start damaging.

3. Common Myna

Classification

Order	:	Passeriformes
Family	:	Saturnidae
Scientific name	:	*Acridotheres tristis* (Linn.)
Common name	:	Common *myna* (Hindi: *Gursali* or *Desi myna*)

Distribution : All India and found up to height of 2700 metres.

Nature of damage : It often damages food grain in the fields and the mandies but seldom enters warehouses. Normally, it feeds on insects, earthworms, fruits, vegetables, kitchen wastes and grains in refuse. Contamination with excreta, feathers etc. are irritating.

Identification

1. Size is between bulbul and pigeon measuring 23-27 cm from beak to tail.
2. It is a perky, well groomed, dark brown bird with blackish head, having a white patch on each wing and characterstic bright yellow bill and legs.
3. Calls are sharp, loud and variable. Normal noise is *keek, keek, kuk, kuk, chur, chur* but alarm calls on seeing any danger is *radio.*

Reproduction and habits : Breeding season is from April to September. In each brood, 4-5 beautiful, glossy, spot free blue eggs are laid. Two or three broods are raised in succession

Management : They can be controlled with strychnine-poisoned baits. Feeding them on *chapatti* soaked in 1 per cent fenthion emulsion can also kill them. Destroying the nests goes a long way in checking their multiplication.

4. Pigeon

Classification

Order	:	Columbiformes
Family	:	Columbidae
Scientific name	:	*Columba livea* Gmelin
Common name	:	Common pigeon, blue rock pigeon (Hindi: *Kabutar*)

Distribution : All India and found up to a height of 3800 metres.

Nature of damage : Mainly granivorous, pigeons eat food grains in grain mandies, godowns, etc. They contaminate the grains with their excreta, feathers etc. they are also responsible for transmission of some diseases and even food poisoning. Found in goods sheds, railway stations, shops, houses and threshing yards, rice mills, flour mills etc.

Identification

1. It is slaty green in colour with glistening metallic green, purple and magenta sheen on the neck and upper breast, two dark bars on wings and one across the tail.
2. It makes a noise like *guturgu, guturgu*
3. Total body length is 33 cm.

Reproduction and habits : It breeds round the year but the peak is from January to May with three to five years in a year. Each brood consists of two elongated white unmarked eggs. Male dances by moving around female with puffed neck and making a clear noise of guturgu. The nest is sketchy pad of twigs and straw and is made in old buildings or found occupying shafts of old wells. Normally it stays in flocks and can be seen on the railway plateforms, high walls, electric wires, old buildings sitting in a line. It is well known for being strong flier and known to return to the same place even after covering the long distances or periods.

Management : Repellents are rarely effective. While they could be very easily poisoned, it is exceedingly doubtful if this would be permitted. Destroy eggs and nests during March-June and September-October.

5. Parrot

Classification

Order	:	Psittaciformes
Family	:	Psittacidae
Scientific name	:	*Psitticula krameri* Scopoli
Common name	:	Rose ringed parakeet (Hindi: *Tota*)

Distribution : Common throughout India and found at heights below 1200 metres.

Nature of damage : Normally frugivorous, it also attacks ripening cereal crops and food grains in open storage. The damage is more than what they actually consume.

Identification

1. It is parakeet green in colour with typical short, stout, deeply hooked red bill and black and rose pink ring round the collar. The female has no collar ring.
2. It has a fine pointed tail, short legs and climbing feet. It is one of the ornamental birds.
3. Some species are good mimics. Makes noise crying *keeak keeak* while flying or at rest.

Reproduction and habits **:** A female lays 4-6 eggs in one brood, which are pure white roundish and oval. The parrots fly in large flocks. The nests are shabby and made in the grooves of the trees or building in a colonial way.

6. Red Vented Bulbul

Order : Passeriformes
Family : Pycnonotidae
Scientific name : *Pycnonotus cafer* Linn.
Common name : Red vented bulbul (Hindi: *Bulbul*)

Distribution : Throughout India

Nature of damage : The red- vented bulbul is used as a fighting bird. This is a pest of garden crops, particularly peas, fruits and flowers. This is also a pest of shops where food is displayed. The pest is well distributed throughout India.

Identification : This bird can be identified with its red coloured vent. This is a medium sized, bird.

Reproduction and habits : Breeding dates are February to August and there are apparently two broods. Incubation period is 14 days. The nests are made up of grass fibres and cobwebs in bushes or trees. Eggs 2 to 4, pinkish white, blotched with dark red- brown. Both sexes take care of the young ones. This is a common garden bird associated with man.

7. Baya Weaver

Order	:	Passeriformes
Family	:	Ploceidae
Scientific name	:	*Ploceus philippinus* (Linn.)
Common name	:	*Baya*

Distribution : The common weaver-bird is distributed throughout India, Pakistan, Myanmar and Srilanka.

Nature of damage : The birds are active during the paddy season because this crop provides food as well as nesting material. In paddy areas, they roost in very large numbers as governed by their seasonal local migration. They can cause quite heavy damage, particularly to the early ripening crop.

Identification: It is of the size of house sparrow and colour resembles the female sparrow when it is not in the breeding season. It is dark streaked, fulvous brown above and plain white fulvous below.

Reproduction and habits : Colonial birds, roosting in enormous numbers near tanks. The male builds a number of intricately and compactly woven nests which are retort shaped with a long vertical entrance tube. The nests are made up of paddy leaves and grasses and are seen in clusters hanging from *ber* or other medium sized trees near ponds and water holes. Mud is used for plastering the egg chamber inside. By completing these nests, the male attracts a number of females who then incubate their respective clutches of 2-4 pure white eggs. Breeding takes place from May to September.

Management : The periodic collection and destruction of nests would lead to reduction in their population.

Crow

Sparrow

Pigeon

Parrot

Red vented Bulbul

Baya weaver

Fig. 4: Avian Pests

Chapter 7

Rodents and Their Management

Economic Importance of Rodent Pests

- They consume and damage foods in the fields and in stores. In addition, they spoil it in store by urine and droppings reducing the sale values.
- Due to their gnawing and burrowing habit they destroy many articles (packing, clothes and furniture) and structures (floors, buildings, bridges, etc). By gnawing the electrical cables they can cause fires.
- They are responsible for transmitting diseases dangerous to human and domestic animals.

Among many pests of agricultural crops and stored grain products, rodents are the major pests which do the great damage. Rodents including rats, mice and squirrels are gnawing animals having chisel like teeth. They have highly developed sense of smell, taste and hearing. They are omnivorous and feed on grains, vegetables, fruits, meat and other products in the houses or in the fields. Rats daily consume food equal to about 10 per cent of their body weight and the figure may go up to 30 per cent in case of mice. Rats are among the most destructive pests of paddy in the field causing about 5-10 percent damage. It is estimated that these animals destroy 30 per cent of the total world's cereal production. In India, 10 million tonnes of food grains are lost every year. It is often said that if rodent ravages could be controlled, a huge amount of food grains can be saved.

Of the 84 species of order Rodentia in India, 50 belong to family Muridae. The important ones belong to two sub-families, viz., Murinae and Gerbillinae. Murinae has 18 species of genus *Rattus*, 2 of *Bandicota*, 7 of *Mus*, 2 of *Apodemus* and a few each under eight other genera. The sub-family Gerbillinae has 2 species of *Gerbillus*, one each of *Tatera* and *Meriones*. Family Sciuridae has 10 species of flying squirrels and 13 species of tree squirrels in India.

Systematic Position of Rat in Animal Kingdom

Phylum : Chordata
Sub phylum : Vertebrata
Division : Gnathostomata
Super- class : Teterapoda
Class : Mammalia
Sub- class : Theria
Infra- class : Eutheria
Order : Rodentia
Family : Muridae

The rats can be divided into two main groups:

1. House Rats

(a) House rat : *Rattus rattus* Linn.
(b) Brown rat : *R. norvegicus* Erxleben
(c) House mouse : *Mus musculus* Linn.

2. Field Rats

(a) Indian mole rat : *Bandicota bengalensis* ray
(b) Large bandicoot rat : *B. indica* Bechstein
(c) Indian gerbil : *Tatera indica* Hardwicke
(d) Soft furred rat : *Rattus meltada* (Gray)
(e) Field mouse : *Mus boodaga* Gray
(f) Brown spiny mouse : *Mus platythrix*
(g) Bush rat : *Golunda ellioti* Gray
(h) Short tailed bandicoot : *Nesokia indica* Gray

(i) Indian desert gerbil : *Meriones hurrianae* (Jerdon)
(j) Pacific rat : *Rattus exulans*

3. Squirrel

(a) Five striped : *Funambulus pennanti* Wroughton

(b) Three-striped : *F. palmarum* Linn.

General Characters of Common Rat Species

1. House Rat

Scientific Name : *Rattus rattus*

Hindi name : *Chuha*

Distribution: It is the commonest house rat distributed all over India and the world.

Characters

(a) Grey to black coloured hair in belly are rough.

(b) It has small eyes, large hair on ears.

(c) Snout is pointed.

(d) Tail is thin uniformly dark coloured.

(e) Weight of adult is 150- 200 g.

(f) Generally, droppings are found scattered and female is banana shaped.

(g) Female has 10 mamme.

Habits

While these prefer to live with man, they can establish themselves in the fields, preferably near human habitation. It is nocturnal in habit, does not burrow but can make nest on trees. They can also be seen during day time.

Nature of Damage

It eats all food material and can damage wood, plastic rubber and soft material also. It is responsible for plague.

2. Brown Rat or Sheep Rat

Scientific Name : *Rattus norvegicus* Erxleben

Hindi name: *Videsi Chuha*

Distribution: Mainly confined to the major cities nearby sea in India and some larger inland cities where it has been introduced. This species has been nowhere as successfull in India as in other parts of the world. It occurs from Arctic to tropic zones but is essentially a temperate zone species.

Characters

(a) Soft skinned with large light brown and whitish belly.

(b) More fleshy than *Rattus rattus.*

(c) Weight is 200-300 gm. Total length 35-41 cm, tail 17-19 cm. Tail is uniformly appeared shorter than head and body length.

(d) Rings on tail 167- 170, faintly marked.

(e) Pear are small with thick furs.

(f) Dropping found in groups and spindle shaped. Weight of dropping is 0.081 gm and mammae are 12 in number.

Nature of Damage

It is omnivorous and also feeds on grain, damaged container, *i.e.* bags carton and spoiled with dropping and hair.

3. House Mouse

Scientific name: *Mus musculus* Linn.

Distribution: Mus musculus may have originally been distributed from the Mediterranean region to China but it has now been spread throughout the world by human and lives as a human commensal.

Characters

(a) House mice are from 65 to 95 mm long from the tip of their nose to the end of their body.

(b) Their tails are 60 to 105 mm long.

(c) Their fur ranges in colour from light brown to black and they generally have white or buffy bellys.

(d) They have long tails that have very little fur and have circular rows of scales (annulations).

(e) Weight ranges from 12 to 30 g.

(f) Many domestic forms of mice have been developed that vary in colour from white to black with spots.

Habits

House mice generally live in close association with human in houses, barns, granaries etc. They also occupy cultivated fields, fence rows and even wooded areas but they seldom stray far from buildings. Some individuals spend the summer in fields and move into barns and houses with the onset of cool autumn weather. Because of their association with human, house mice have been able to inhabit inhospitable areas which they would not be able to occupy independently.

Nature of Damage

In the wild, house mice eat many kinds of plant matter such as seeds, fleshy roots, leaves and stems. Insects and meat (carrion) may be taken when available. In human habitation, *Mus musculus* consumes any human food that is accessible as well as glue, soap, and other household materials. Many mice store their food or live within a human food storage facility.

4. Large Bandicoot

Scientific Name : *Bandicota indica* Bechstein

Hindi name : *Ghoos*

Distribution : Rajasthan to penisular India, Bihar, orissa Assam and Nepal.

Characters

(a) It is largest field rat in size than other species of rats.

(b) Mostly, it is found in blackish or blackish brown in colour.

(c) Head is broad with long whiskers.

(d) Head and body length is 51- 68 mm.

(e) Ears short and round.

(f) Eyes are small with white eye brow.

(g) Ears are thick coarse with spines.

(h) Weight is about 800- 1000 gms.

(i) Tail is equal to length of the head and body.

(j) Rings of tails are 230- 240 and they are not clear.

(k) Droppings are scattered and spindle shaped weight about 0.064 gm.

(l) Fore limbs are smaller than hind limbs and mammae are about 18 in number.

(m) It contains 2-5 opening. They are mainly grain eaters.

Habits

It is large house mouce generally live in close association with human in houses, barns, granaries, etc. They also occupy cultivated fields, fence rows and even wooded areas, but they seldom stray far from buildings. Some individuals spend the summer in fields and move into barns and houses with the onset of cool autumn weather. Because of their association with human, house mouce has been able to inhabit inhospitable areas which they would not be able to occupy independently.

Nature of Damage

It makes big burrow which weakens the foundation, walls, rivers, bunds and railway tracks. It damages crop by tillers and leaves reducing yield drastically.

5. Indian Mole Rat

Scientific Name : *Bandicota bengalensis* Gray

Hindi name : *Chuha Ghoos*

Distribution : All-India

Characters

(a) Body is robust.

(b) Colour is dark brown to black.

(c) Head and body length is about 25-35, cm

(d) Head is short and snout is pig like.

(e) Tail is short having length of about 10 – 20 cm. Slightly paler yellow in colour from the lower side.

(f) Weight approximately 326 g (male) and 287 g (female)

(g) Ring on the tails are 160-170 clearly seen.

(h) Dropping is scattered and oval shaped, weight about 0.042 gm.

(i) Ears are big having length of about 2.5 cm and without hairs.

(j) Furs are short rough and spine like in structure.

(k) Mammae are 12-18 in number.

Habits

Fossorial animal wherever possible but extremely adaptable. Nocturnal normally, when population is dense then also can be seen during the day. Primarily, parasitic on man but can survive on feral populations.

Nature of Damage

It feeds on grains, fruits, vegetables some time feed young ones and dead animals. It damages crop in field like sugarcane, paddy, wheat and maize. They cut earheads, therefore, yield is reduced.

6. Short Tailed Bandicoot

Scientific Name : *Nesokia indica* Gray

Distribution : It covers a wide range from Egypt to Sinkiang and Northern India

Characters

(a) The head and body of *N. indica* measures 140-215 mm and its tail length is 88-129 mm.

(b) Its robust form characterises this rat species.

(c) The head is short and rounded and the muzzle is short and broad.

(d) The rat has eight mammae. The pelage of the bandicoot rat varies from coarse, short, harsh and semispinous to long, fine and silky.

(e) The consistency of the coat has been found to be both dense and thin.

(f) The feet and tail are scantily haired.

(g) The feet are also broad and the ears are rounded.

(h) The incisors of *N. indica* are stout and broad.

Habits

The bandicoot rat lives mainly in moist areas or along streams and canals. It is found in deserts, steppes, cultivated areas, and forests. The elevational range is from about 26 meters below to about 1,500 meters above sea. The bandicoot rat prefers to live in areas with soft soil and good vegetation coverage. When the soil becomes dry, *N. indica* migrates away to softer soil.

Nature of Damage

The rat feeds on leaves and roots. However, it also consumes crops such as barley, wheat, potato, groundnut sugarcane, mustard, brinjal, and water melon. This rat sometimes stores food in its burrow, up to 7.3 kg of wheat ears has been found hoarded in a single burrow.

7. Field mouse : *Mus boodaga* (Gray)

It is found in dry as well as in wet lands and is a fairly good swimmer. It is brown from above and whitish or dull grey on the under side. It measures 2-3 inches with slightly shorter tail. It is nocturnal in habit. It dwells in shallow burrows with one or two openings. The length of the burrows ranges from 50-116 cm and its diameter is 2-3 cm. It feeds on paddy grains and at times steals from the food chambers of *B. bengalensis.*

8. Soft furred rat *: Rattus meltada* (Gray)

This species is found in wet lands as well as in dry lands having many habits similar to those of *B. bengalensis.* It is 5- 6 inches long with equal or shorter tail and about 60 gm in weight. It bears large rounded ear lobes and soft coat of fur. It is mostly found in irrigated land, bunds hedges and grasslands and nocturnal in habit. They are fairly good swimmer and have the characteristic habit of cutting a paddy ear and swimming back to the bund with the booty.

9. Indian gerbil : *Tatera indica* (Hardwicke)

It is also known as the antelope rat and is found in sandy soils. This has 6- 7 inches long tail and reddish- brown or greyish dorsum and white under side. It is nocturnal in habit and lives in simple burrows with 3- 10 openings. The burrow opening is often located under the cover of a thorny bush or some vegetation but the scattered freshly excavated soil helps in its direction. It feeds on seeds in the winter; stems, rhizomes and insects in the summer and on leaves and flowers of plants in the monsoon.

10. Indian desert gerbil : *Mriones hurrianae* (Jerdon)

It inhabitats the arid and semi-arid regions of Punjab, Haryana, Rajasthan and spread up to Pakistan, Afghanistan and Iran. It is a diurnal rat and digs extensive burrows with 10-15 openings. Its body is covered with black tassel of hair at its tip. It causes damage to the crops and bark of the trees.

11. Pacific rat : *Rattus exulans*

This is relatively small species and colour is gray brown dorsally and light grey ventrally. The body length is 100-130 mm and a fully grown adult may weigh up to 45 g. The tail is longer than the body length. It is found in agricultural field and in villages. Due to its excellent climbing ability, it is common pest of coconut trees.

12. Five striped squirrel : *Funambulus pennanti* and Wroughton Three striped squirrel : *F. palmarum* (Linn.)

Distribution: F. pennanti is found in Northern India and *F. palmarum* in Southern India

Habits : The *F. pennanti* is a commensal rodent, parasitic on man, and inhabiting the centres of villages and cities. Palm squirrel of Southern India lives in the forests.

Hosts : Flowers, fruits, nuts of various plants and also insects.

Nature of Damage

Nuisance around houses. As both have been demonstrated to be plague carriers and because of their commensal habits, they must be considered potential pests in this regard.

Habit and Habitat

Good swimmer, expert in digging burrows with characteristic mounds of soil which light of opening burrows. Hence the name is mole rat. Rats are cosmopolitan in their distribution. Rats have highly developed sense of smell, taste and hearing. They are omnivorous and they come out at the night and remain hidden at the day time. They are good swimmer, climber and they can jump horizontally about 8 feet. Generally, rats live in burrows but they can also be found in heap of wood, rubbish etc. One pair of rat can reproduce 1000 of rats in a year.

Life History

A rat can breed after attaining age of 3-4 months. Female rat breeds 6-10 times in a year. Rat can reproduce throughout the year. The gestation period of rat is about 20- 27 days. The field rats breed twice in a year but the house rats breed 4- 6 times in a year giving birth to 8- 10 young ones in a litter. A pair of rat can produce 1000 in a year. The life span of a rat is 2- 3 years.

How to Detect Presence of Rats in the Field

Rats are found in the heap of wood, burrows, or even in the ditch. The depth of a burrow may be 0.3 – 1.4 m. One burrow may contain 2- 42 openings. There are three types of burrows:

(a) *Round type*: The diameter varies from 1.5 to 2.0 inch. Most of these are not inhabited by rats.

(b) *Irregular type*: Burrows which are irregular in shape and the fresh soil is very often found at the base of these burrows. These holes are slanting inside and leads to the burrows and such burrows are inhabited by rats.

(c) *Semi – circular type*: This type of burrows has a diameter of about 2-3 inch. These burrows are inhabited by rat generally by females which live matured stage of pregnancy with the freshly born young ones.

How to Detect Presence of Rats in Warehouses

From following points the presence of rat can be detected in warehouses:

(a) *Spoilage*: Gnawed bags with grain scattered on floor.

(b) *Dropping*: Excreta of rat is known as rat dropping. Dropping indicates the presence of rat.

(c) *Grnawing marks*: Presence of regular holes in wooden doors and windows indicate the presence of rats.

(d) *Run ways*: Marks of foot and tail left on dusty floor represent active rat infestation in godown.

(e) *Holes and scrap*: The presence of holes with scraps and newly dug out loosen soil heaped any where inside or around godown give information of infestation.

(f) *Disappearance of baits*: When we put bait at the active burrows and if it disappears, it is sure that the rats are present or presence may be inspected.

(g) *By noise of rat*: The presence may be suspected.

(h) *Odour :* Rats have little odour unless the population is large or well established. Mice can offen be detected by their musty odour.

Management of Rodents

Why is managing rodents in storage important?

Feeding on Stored Produce

Rats consume about 25 gm of food per day and mouce eats approximately 3-4 gm per day. They contaminate a lot of the stored produce with urine, faeces, hair and pathogenic agents. Infested batches often have to be declared unfit for human consumption. As rodents mainly feed on the embryo, they cause particular damage to the nutritional value and germination ability of seeds.

Damage to Material and Equipment

Rodents damage tarpaulins, bags, pallets, sprayers and to the store itself (cables, doors). These often lead to subsequent damage like produce leaking out of bags or containers, bag stack collapsing, sparks or fire from short circuits, silos and warehouses may subside or even collapse as a result of being undermined and damaged drainage canals around a store.

Planning: A good amount of planning should be done before rodent control operation is started. First of all, the cost worthiness of the control should be calculated and if worthy, the area where control is to be carried out should be surveyed and mapped and manpower needed is estimated. Management of rodents may be in two groups:

(a) Rat prevention

(b) Rat destruction

a. Rat Prevention

(i) Keeping Rodents Out

The rat population is directly proportionate to the availability of things. If we remove these two basic things, then rat damage can be prevented to a great extent. The following points should be kept in mind while construction or hiring a building:

- All store buildings should be *Pucca* it means it must be made up of concrete and plastered with cement.
- Plinth without steps should be constructed high above ground level.
- The wooden door should be tightly fitted with the floor. It will be better if 15 cm of bottom of the door is covered with tin plate.

- Windows and ventilator opening should not be less than 1 m. from ground level. They should be covered by 1 cm wire netting.
- All the points of entry below 1 m. from the ground level is closed.
- All the burrows inside building should be sealed with cement.
- Other source of rat entry like branches of tree, wires and ropes should be removed.
- All the sources of water should be removed around the building.

(ii) Storage Hygiene and Technical Measures

- Keep the store absolutely clean. Remove any spilt grain immediately.
- Store bags in tidy stacks set up on pallets, ensuring that there is a space of 1 m all round the stack.
- Store any empty or old bags and fumigation sheets on pallets, and if possible in separate stores.
- Keep the store free of rubbish in order not to provide the pests with any places to hide or nest. Burn or bury it.
- Keep the area surrounding the store free of tall weeds so as not to give the animals any cover. They have an aversion to crossing open spaces.
- Keep the area in the vicinity of the store free of any stagnant water and ensure that rain water is drained away, as it can be used as source of drinking water

b. Rat Destruction

It includes

(a) Mechanical control

(b) Chemical control

(i) Mechanical Control

(a) *Hunting*: The rats can be killed with the help of trained cats and dog etc. Normally predator will not keep rats and mice at economical population levels. Cats do not directly control rats and mice by feeding them but their presence keeping them away.

(b) *Trapping*: Trapping is suitable for small population like that of house mouse (*M. musculus*). The rat can be collected and killed with the help of wooden and iron trap by placing a suitable bait inside the trap and can be utilized as a food in the area of Nagaland. This method has one serious disadvantage that only some individual enter the trap.

(c) *Glue Boards*: One of the alternatives to a snap trap is a glue board. Glue boards work on the same principle as fly paper: when a rat or mouse attempts to cross the glue board, the rodent gets stucked. Glue boards are much more effective for mice than for rats. Also, one of the major drawbacks with glue boards (and other live-catch type traps) is that the trapped rat may not die quickly and we have to kill it. For this reason, glue boards are not good alternatives for many people and their use is not recommended. Also, cats and dogs may get into the glue and track it around the house, creating additional problems.

(d) *Electrocution trap :* These traps are baited with a suitable bait (peanut butter, cheese, oats, salami, nesting materials etc.) and placed in the rats travel paths. When a rat enters the trap, it immediately receives a lethal electric shock. The traps that kill rats by electrocution are available in the markets. These traps are considerably more expensive than the common trap and can be used safely and effectively only in limited situation.

(ii) Chemical Control

The chemical which kills the rats are known as rodenticide or raticide. They are commonly used in the form of poison bait. The poison used in the baits are of two types: (a) Acute poison, and (b) Chronic poison.

(a) *Acute poisons* : These are the poisons which are used in a single dose or these are the chemical which kill the rat by single dose, *e.g.* zinc phosphide, bromethalin, cholecalciferol etc.

(b) *Chronic poisons*: These poisons are lethal when consumed for several days as they cause external and internal haemorrhage, *e.g.* warfarin, chlorophacinose, diphacinonene, brodifacoum, difethialone etc.

Toxic Bait

While trapping is generally recommended for controlling rats indoors, when the number of rats around a building is high, we may need to use toxic baits to achieve adequate control, especially if there is a continuous reinfestation from surrounding areas. If this is the case, consider hiring a licensed pest control applicator, who is trained to use rodenticides safely.

Toxic baits contain active ingredients that work as either an anticoagulant, causing death by internal bleeding or as a direct toxin (Table 7.1).

Table 7.1: Types of baits used in rat control & their active ingredients.

Type of bait	Active ingredients	Use restrictions
Multiple-feeding anticoagulant	Warfarin, Chlorophacinone, Diphacinonene	For use in and around buildings and in gardens and landscaped areas
Single-feeding anticoagulant	Brodifacoum, Bromadiolone, Difethialone	For use in and around buildings only
Non anticoagulant	Bromethalin, Cholecalciferol	For use in and around buildings only
Non-anticoagulant	Zinc phosphide	For use in and around buildings and in gardens and land scaped areas

a. Non-anticoagulant Rodenticides

The older rodenticides, formally referred to as the acute toxicants (*e.g.*, arsenic, red squill and phosphorus) are either no longer registered or of little importance in commensal rodent control. Newer rodenticides are much more effective and have resulted in phasing out of these older materials over the last 20 years.

At present there are three non-anticoagulant rodenti-cides: zinc phosphide, cholecalciferol (Vitamin D_3) and bromethalin, registered and available for commensal rodent control. Since, none of these are anticoagulants, all can be used to control anticoagulant resistant rodent populations.

Of these active ingredients, bromethalin and cholecalciferol are formulated to serve as chronic rodenticides, applied so that rodents will have the opportunity to feed on the baits one or more times over the period of one to several days. Because they are slow-acting in comparison to zinc phosphide, bait shyness is neither a problem nor is prebaiting is necessary to get good control in most situations. Zinc phosphide differs in that prebaiting (offering rodents similar but non-toxic bait prior to applying the toxicant-treated bait) is recommended to increase bait acceptance. Zinc phosphide is not designed to be left available to rodents for more than a few days, as continued exposure is likely to result in bait shyness within the population.

Non-anticoagulant rodenticides, particularly, zinc phosphide remain useful tools to achieve rapid reductions in rodent populations. When population levels are high, the cost of baiting with these materials may be lower than the anticoagulants.

(a) *Bromethalin:* (Trade name: Assault, Vengeance, Trounce) N-methyl-2, 4-dinitro-N- (2, 4, 6-tribromo-phenyl)-6-(trifluoromethyl) benzenamine

Bromethalin is a single-dose rodenticide that causes central nervous system depression and paralysis, leading to death in 2 to 4 days. Bait should be renewed at interval of several days. Continuous bait availability (as with anticoagulants) is not required but bait needs to be presenting long enough to allow all animals in the area to feed. The amount of bait needed is usually about one-third of that used with anticoagulants, since an animal ingesting a lethal dose does not feed again. This effect is unlike that of anticoagulants in which rodents continue to consume bait after they have ingested a lethal dose. Bait shyness has not been reported.

(b) *Cholecalciferol* : (Trade name: Vitamin D_3, Quintox, Rampage) 9, 10-Seocholesta-5, 7,10 (19)-trein-3 betaol.

Cholecalciferol is a single-dose or multiple-dose rodenticide that causes mobilization of calcium from the bone matrix to plasma and death from hypercalcemia. Time of death is 3 to 4 days after ingestion of a lethal dose. As the toxicant is slow-acting, bait shyness apparently does not occur. As with Bromethalin, once a rodent consumes a lethal dose, all food intake ceases.

(c) *Zinc phosphide*: Zinc phosphide is a dark grey powder, insoluble in water that has been used extensively in the control of rodents. When zinc phosphide comes into contact with dilute acids in the stomach, phosphine (PH_3) is released. It is this substance that probably causes death. Rats and mice that ingest lethal amounts of bait usually succumb overnight with terminal symptoms of convulsions, paralysis, coma, and death from asphyxia. They typically die in a prone position with their legs and tails outstretched. Because zinc phosphide is not stored in muscle or other tissues of poisoned animals, there is no secondary poisoning with this rodenticide. The bait, however, remains toxic up to several days in the gut of a dead rodent. Other animals can be poisoned if they eat enough of the gut content of rodents recently killed with zinc phosphide.

Zinc phosphide is available in ready-to-use dry baits and also in concentrates for use by persons trained in rodent control who may wish to prepare their own baits. Its strong garlic-like odour appears to be attractive to rodents that are not bait-shy and apparently makes the bait unattractive to some other animals. Bait shyness can be a problem; so prebaiting is recommended or necessary for achieving good bait acceptance.

b. Anticoagulants

(a) *Warfarin* : Its chemical name is (alpha-acetonylbenzyl)-4-hydroxycoumarin. It was the first marketed anticoagulant and, therefore, became the best known and most widely used. It has relatively limited sales today, due to the availability of more potent anticoagulants.

(b) *Chlorophacinone* : (Trade name: RoZol). Chemical name: 2-[(p-chlorophenyl)phenylacetyl]-1,3-indandione.

(c) *Diphacinone* : (Trade name: Ramik, Ditrac). Chemical name: 2-diphenylacetyl-1,3-indandione.

Chlorophacinone and diphacinone are similar in potency and are significantly more toxic than the anticoagulant compounds developed earlier. Consequently, they are formulated at lower concentrations. Chlorophacinone and diphacinone may kill some rodents in a single feeding but multiple feedings are needed to give adequate control of an entire population.

(d) *Brodifacoum* : (Tradename: Talon, Havoc). Chemical name: 3-[3-(4'bromo[1,1'-biphenyl]-4-yl)-1,2,3,4,-tetrahydro-1-naphalenyl]-4-hydroxy-2H-1-benzophyran-2-one. Brodifacoum is the most potent rodenticide currently available for commensal rodents. It is available in 0.005% pellet formulations and in wax blocks. Because of its acute toxicity, a lethal dose can be obtained in a single feeding, although death is delayed for 4 or 5 days.

(e) *Bromadiolone* : (Tradename: Maki, Contrac). Chemical name: 3-[3-(4 bromo[1,1 biphenyl]-4-yl)-3-hydroxy-1-phenylpropyl]-4-hydroxy-2H-1-benzopyran-2-one.

Bromadiolone is not quite as toxic to rodents as brodifacoum but can result in the same level of control. It is available in 0.005% pellet formulations and in wax blocks.

Technique of Baiting

Bait is the edible material, which is used to attack the rat and to feed upon it. Bait may be poisonous or non poisonous. Baiting is required in case of chemical control. There are two types of baiting: (a) Surplus pre-baiting and (b) Token pre-baiting.

In case surplus pre-baiting, rats are kept half fed. It is well known that token pre-baiting is better than surplus pre-baiting. The bait should either be kept scattered or put in small heaps at various spots in the godowns during night. In the field conditions, the bait should be introduced deep in burrows with the help of a wooden spoon at a rate of 25-30 g per burrow depending upon the size. It has been found advantageous to do the pre-baiting on days one and three and poison baiting on the fifth day on days two, four and six, another area may be covered in the same way. To test the success of poison baiting, plain baits or non poisoned baits should be placed after poison baiting, if the baits have been eaten by rats, it indicate that complete success has not been achieved and poison baiting should be repeated again.

Composition and Preparation of Different Baits

1. Zinc Phosphide

The most commonly used and effective acute poison is zinc phosphide. It is a dark grey coloured powder and continued use causes bait shyness.

Composition of bait

(a) Zinc phosphide : 2 parts

(b) Food material : 97 parts

(c) Vegetable oil : 1 part

Wheat, bajra, barley or gram flour can be used as a good material or non poisoned bait.

2. Strychnine

It is a highly poisonous chemical. The bait is prepared in the following manner: (a) 27 g strychnine is dissolved in 56 cc of water. (b) 2 kg of gur is mixed in 455 cc of warm water.

Now mixture (a) and (b) are mixed. Now we will mix 13 kg of over night soaked seeds of gram in the above mixture. Now, we can put 24 g or 20 ml of this poison bait at every active burrow.

3. Warfarin

It is a good chornic poison and does not require pre baiting. The following ingredients are used for the preparation of poison bait:

(a) Warfarin – 25 g

(b) Ground grain – 450 g

(c) Sugar – 15 g

(d) Edible oil – 10 g

4. Prolin

Rats some time do not die after eating warfarin because a bacterium develop in their stomach and this bacterium is responsible for preparing a vitamin which acts as a antidote against warfarin, that is why the warfarin does not act. When warfarin is mixed with sulfaquinone, it acts as a good poison which is known as prolin. After mixing the sulfaquinone in warfarin, the bacterium dies in the stomach of rat, that is why the prolin act as a good poison.

5. Barium Carbonate

It causes rapid bait shyness very toxic to mammal, less and slow acting poison, therefore, it is not popular.

(a) Barium carbonates – 16 parts

(b) Food bait – 83 parts

(c) Vegetable oil – 1 part

6. Red Squill

This is a tasteless poison so rat eat more and more.

Fumigation : The control of rodents by fumigation can be very effective but it may be expensive and dangerous. It should be remembered that the gas must have access to burrows, if these are present in the building, *i.e.* the burrows should be open and the fumigant used must be heavier than air. If the species concerned makes burrows which are easy to locate, they can be fumigated directly. The simplest method is to use a powder which releases hydrogen cyanide or aluminum phosphide tablets which release phosphine when placed into the burrows. The gases are generated when the powder or tablets are in contact with moisture in the field. Alternately, methyl bromide gas may be pumped into the burrow system.

As soon as fumigant has been applied, all burrows must be closed by filling the entrance holes with earth. However, fumigants cannot be used in loose or sandy soil as too much gas escapes and the treatment may not be effective. Occasionally, rats have been known to block tunnels and prevent complete distribution of the gas so some individuals survive. It is important, therefore, always to check reopened burrows or other signs of survival a few days after the prescribed fumigation period.

Chapter 8

Micro Organisms

As the name indicates, micro-organisms are very minute, cellular, living organisms which cannot be seen with the naked eyes (however, large colonies of fungi can be seen). Micro-organisms affecting stored grain can be divided into the following groups in order of their significance:

(i) Fungi

(ii) Actinomycetes

(iii) Yeast

(iv) Bacteria

(v) Protozoa (not found in sound food grains)

Fungi, bacteria and actinomycetes are grouped under the Thallophyta of the plant kingdom. The most important factors for their normal development are: (i) Nature of Food, (ii) Moisture, (iii) Oxygen and (iv) Temperature.

Fungi

Characteristics of Fungi

These are achlorophyllous, slender, fine branched thread like, filamentous structures connected with each other forming mycelia. Protoplasmic activity at the tips of hyphae is highest where growth manifests itself into apical elongation behind which are cross walls and reproductive structures. The reproduction of fungi is through spores of either sexual or asexual nature. Sexual spores are formed after the fusion of two cells and asexual spores are formed without any fusion of cells. Spores may be unicellular or multicellular of many shapes. The spores vary in length of time in their viability. The set of factors which constitute a favourable environment for the development of spores are as follows:

(a) Moisture content above 12 per cent.

(b) Temperature 30-32°C.

(c) The degree to which the grain has been invaded by fungi or insects before or after storage.

(d) The amount of foreign matter (dust, dirt, chaff, straw and other organic matters).

Viable spores germinate and produce one or more hyphae which invade grain and grain products. The presence of toxic substances, other microbes, mechanical obstruction etc. hinder their progress. Fungi affecting food grains can be divided into two groups:

(i) *Field fungi* : The field fungi invade seeds either during development or after maturity but before harvest. Generally, field fungi cause damage in the field and no damage or little damage during in storage. These fungi cause damage to seeds, when the moisture content is in equilibrium with relative humidities above 95 per cent, which gives a seed moisture content of 24 to 25 per cent on a wet basis in seeds of cereals. The moisture content is far above that at which these seeds are stored, hence field fungi rarely cause damage during storage, *e.g. Alternaria* spp., *Cladosporium* spp., *Hetminthosporium* spp., *Fusarium* spp. etc.

(ii) *Storage fungi* : Storage fungi are those that grow in seeds when the moisture content is in equilibrium with 65 to 90 per cent relative humidities. Stoage fungi normally do not play a role in disease development in the field but play a major role in seed deterioration in storage. Main storage fungi are species of *Aspergillus*, such as *A. amstelodami, A.*

candidus, A. chevalieri, A. flavus, A. repens, A. restrictus, A. rubber and *A. ochraceus. Penicillium,* which usually invades seed having moisture content above 16 per cent and at low temperature. *P. virdicatum* is more common in temperate than in tropical regions. Species of *Sporendonema, Mucor, Rhizopus, Absidia, Byssochlamys* and *Hansenula* have also been reported on stored seeds.

Detection of Storage Fungi

Fungi found externally on grain can be detected by the following methods :-

(a) Surface detection of growth when the stocks are stored moist.

(b) By examining the infected seedlings.

(c) Examination of grain tissues under microscope.

(d) Examination of discoloured, damaged shriveled embryo or grain.

(e) Aqueous washings of grains.

(f) By culturing the washing etc.

(g) By fat acidity values.

(h) Increase of temperature in the grain (up to 55°C).

Invasion by Storage Fungi

Species of *Aspergillus* and *Penicillium* normally do not infect prior to storage but invade under favourable conditions prevailing in storage. A storage fungi enters seed tissues through injuries to the pericarp on testa. In addition, maturing seed can be damaged on the parent plant by birds or insects, and fungi can readily infect these damaged tissues. *A. flavus* var. *columnaris* may enter to seed tissues *via* micropyle and peduncal or penetrate through microscopic cracks in seed surface of maize. The peduncle scar tissue is loose parenchyma, which allows passage for advancing fungal mycelium.

Losses Due to Fungi

(i) *Reduction in germination*: Reduction in germination can be caused by mechanical or physiological or by storage fungi. If caused by fungi, the effect is influenceds by seed moisture content, storage period, storage conditions (temperature and humidity) and other factors. Decrease in germination and seedling vigour results from ageing or infection by storage fungi. Seed deterioration occurs at faster rate when temperature and moisture contents are favourable for fungi.

(ii) *Discolouration and shrinkage of seed*: *Tilletia indica*, Karnal bunt of wheat cause black discolouration at the embryonal end and groove of the seed. *Alternaria alternata*, black point of wheat cause brown to dark brown discolouration on the part of seed or whole seed. *Aspergillus* and *Penicillium* cause discolouration and shrinkage of barley seeds.

(iii) *Production of heat* : High temperature in storage increases the respiration rate which lead to loss in weight and quality of seed. Cereal and legume seeds have low heat conductivity, so that the effects of temperature fluctuations are noticeable over short distances or longer periods. This results in heat accumulation in pockets. Heat damage may also be due to fungal activity and not heat, although both factors are present when stored seed or grain undergoes heating due to fungal growth.

(iv) *Reduction in nutritional value:* The storage fungi break down the nutrients of seeds chemically.

   (a) *Increase in fatty acid value*: Seeds contain lipids and when seeds are damaged by improper storage conditions or by microorganism, lipid generation occurs. Lipase and lipoxygenase are the principal enzymes involved in lipid generation in seeds. Lipase may arise through activation of pre-exiting materials, denovo synthesis or through production by micro-organisms. In each case, the mechanism is similar and the end result is the same, an increase in free fatty acid, which is an important parameter of oilseed quality. According to the National Cottonseed Products Association Trading Rules, grade No. 1, sunflower seed shall contain not more than 1.8 per cent free fatty acid in the oil and seeds and seeds containing over 3 per cent free fatty acid shall be sample grade and subject to rejection.

(b) *Biochemical alteration in nutritional value*: Microorganisms grow luxuriantly on seeds and decompose proteins to lower molecular weight components, resulting in the formation of free amino acids. Moreover, there is deletion of some enzymes and an intensification of others and/or production of new multiple molecular forms of enzymes. Many of these enzymes in seeds remain active during the infection period.

(v) *Production of toxins*: Storage fungi may produce mycotoxins that are injurious to man and animals on consumption.

(a) *Aflatoxins*: *Aspergillus flavus* produces aflatoxin B_1 and B_2, and *A. parasiticus* aflatoxin G_1, G_2, M_1 as well as B_1 and B_2. Aflatoxin B_1 causes liver damage to most mammals and even human beings. Aflatoxins caused human fatalities in the Third World War and under famine conditions when heavily contaminated maize and rice were consumed. Diseases such as hepatitis, heptocarinoma and Reye's syndrome are caused by consumption of aflatoxin contaminated food. Aflatoxins interfere with immune system in certain animals and are carcinogenic and mutagenic. The International Agency for Research on Cancer placed aflatoxin on its list of human carcinogens.

(b) *Ochratoxin A (OA)*: Ochratoxin A is produced by a number of *Aspergillus* and *Penicillium*. In human beings, it can cause degeneration of kidney, which can lead to death in extreme cases.

(c) *Citrinin* : It is produced by *Pencillium citrinum, P. verrucosum* and *Aspergillus terreus.* Citronin is a nephrotoxin that occurs in apple juice, barley, oats, peanuts, rye and wheat. Human kidney damage occurs with prolonged ingestion. It is also toxic to birds, dogs and pigs and it results in watery diarrhoea, increased food consumption due to kidney degeneration.

(d) *Ergot alkaloids*: Ergot alkaloids are produced by *Claviceps* in rye and triticale. Its consumption may result in gangrene, leading to necrosis of extremities, central nervous system effects ataxia, convulsions and paralysis, and gastrointestinal disorders.

(e) *Zearalenone*: It is produced by *Fusarium roseum* and *F. tricinetum*. It results in estrogenic and anabolic activity in animals and is associated with reddening and swelling of the vulva, uterine enlargement and vaginal and rectal prolapse. It occurs in cereals, primarily in maize and animal feeds.

(f) *Citreovirdin*: *Pencillium ochrosalmoneum* and *P. citreonigrum* produce it. The disease beriberi has been associated with consumption of mouldy 'yellow rice'. In Japan in 1910, yellow rice was banned due to this disease. In several animal species, it causes vomiting, convulsion, ascending paralysis and repiratory arrest.

Management

Moisture is one of the important factors which is essential for growth of fungi. The inherent moisture content of grain if reduced to safe level, the chances of fungal growth are minimized. The safe levels of moisture content for wheat and rice are 12 per cent and 13 per cent, respectively. Moisture of food grains can be reduced either by sun drying or with help of grain driers. Restricted aeration also retards the fungal growth but does not prevent the growth of fungi totally. Reduction of temperature to a range of 23-30°C reduces growth rate of a number of species of fungi. For seed treatment, an ideal chemical must have low mammalian toxicity and long lasting microbial inhibiting properties and should not have adverse effects on germination.

Actinomycetes

Following are the common harmful Actinomycetes found in food grains:

(a) *Actinomycetes allido*

(b) *A. globisporus*

(c) *A. graminis*

(d) *A. grisens*

The damage done by bacteria, yeast and actinomycetes are similar but are much less as compared to fungi. The nature of damage caused by them is also similar and so also the control measures.

In short, the best ways to avoid damage from micro-organism during storage of food grains are:

(a) Clean the food grains before storage.

(b) Proper drying of food grains before storage.

(c) Store food grains in receptacles where moisture increase is minimum during rainy season. If it is not possible to do so, aerate the godowns on sunny days.

(d) Cool grain stores well for a longer period.

Yeast

Some common species of harmful yeasts in food grains are:

(a) *Monilia* sp.

(b) *Candida* sp.

(c) *Fusidiunt* sp.

(d) *Torula* sp.

Bacteria

Bacteria are also harmful to food grains. Some important harmful species found in grains are:

(a) *Pseudomonas* spp.

(b) *Salmonella* spp.

Chapter 9

Stored Grain Insect Pests of Quarantine Significance for India

In the past several insect pests had entered India and are causing economically perceptible damage to native crops. The Government of India, in line with the International Standards on Phytosanitary Measures (ISPM), notified 'Plant Quarantine Order, 2003' to reduce the risk of introduction of exotic insect pests in India. The storage insect pests of quarantine concern for India have been listed in Table 9.1.

Table 9.1: Insect species of quarantine importance

S.No.	Host	Scientific Name	Common Name
1	*Cereals*	*Prostephanus truncates*	Larger grain borer
2	*Cereals*	*Sitophilus granarius*	Granary weevil
3	*Cereals*	*Sitophilus zeamais*	Maize weevil
4	*Cereals*	*Trogoderma variable*	Grain dermestid
5	*Coco*	*Conopomorphaa cramerella*	Cocoa moth
6	*Coco*	*Hypothenemus obscurus*	Tropical nut borer
7	*Coco*	*Acorocercops cramerella*	Coffee moth
8	*Coco*	*Sahlbergia singularis*	Mirid
9	*Coco*	*Sahlbergiella theobroma*	Cocoa capsid
10	*Coffee*	*Hypothenemus hampei*	Coffee berry borer
11	*Oil seed*	*Stator pruininus*	Seed bruchid
12	*Oil seed*	*Trogoderma versicolor versicolor*	Dermestid beetle
13	*Pulses*	*Acanthoscelides obtectus*	Bean bruchid
14	*Pulses*	*Anomala cupripes*	Large green chafer beetle
15	*Pulses*	*Anomala pallida*	Large green chafer beetle
16	*Pulses*	*Apomyelois ceratoniae*	Carob moth

(Cont.)

S.No.	Host	Scientific Name	Common Name
17	*Pulses*	*Brucllididillm spp.*	Seed Bruchid
18	*Pulses*	*Bruchus lentis*	Bruchid
19	*Pulses*	*Melanagromyza chalcosoma*	Pod fly
20	*Pulses*	*Specularis impressithorax*	Bruchid
21	*Pulses*	*Spermophagus spp.*	Seed bruchids
22	*Sheanut*	*Ephestia elutella*	Chocolate moth

Insect Species of Quarantine Importance

The mark of identification and quarantine measures for the stored grain insect pests of concern to India have been discussed below:

1. Larger grain borer: *Prostephanus truncates* (Coleoptera: Bostrichidae)

The adult is 3-4.5 mm long with typical cylindrical shape. The declivity is flattened and steep and has many small tubercles over its surface. The limits of the declivity, epically and laterally, are marked by a carina. Antennae are 10 segmented, slender and clothed with long hairs.

Quarantine Measures

- Import of maize seeds for sowing should be free from this pest and subject to prior approval of the Department of Agriculture and Cooperation (DAC).
- Import of maize grains for consumption shall accompany with Phytosanitary Certificate with an additional declaration for the freedom from *P. truncates* and with special condition that consignment is fumigated with methyl bromide @ 32 gm/m^3 for 24 hrs. at 28°C.

2. Granary weevil : *Sitophilus granarius* (L.) (Coleoptera: Curculionidae)

Adults are uniformly polished chestnut-brown or reddish-brown to shiny black in colour. They are 2.5-5.0 mm long having the characteristic rostrum and elbowed antennae. The prothorax has distinctly oval shaped punctures. Adults do not have hind wings, therefore, cannot fly.

Quarantine Measures

- Import of wheat, rice and barley seeds for sowing should be free from this pest and subject to the approval of Department of Agriculture and cooperation.
- Consignment of rice seeds for sowing should be soaked overnight and treated in hot water at 52°C for 10 minutes subject to supervision, monitoring and testing by Director, National Beurau of Plant Genetic Resources, New Delhi or Project Director, Directorate of Rice Research, Hyderabad.
- The consignment should be fumigated with methyl bromide @ 32 gm/m^3 for 24 hrs. at 28°C.

3. Maize weevil : *Sitophilus zeamais* Motschulsky (Coleoptera: Curculionidae)

Usually more black than *S. oryzae* females with lateral lobes of the Y-shaped sclerite pointed and their separation is greater than S. *oryzae*. This species are differentiated from *S. granarius* by the presence of wings beneath the elytra and having circular, rather than oval, shaped punctures on the prothorax.

Quarantine Measures

- Import of maize seeds for sowing should be free from *S.zeamais* and subject to prior approval of the DAC.
- Import of maize grain for consumption shall accompany PSC with an additional declaration of freedom from S. *zeamais* and with special condition that the consignment is fumigated with methyl bromide @ 32 gm/m^3 for 24 hrs at 25°C.

4. Grain dermestid : *Trogoderma variable* (Coleoptera : Dermestidae)

The adults resemble *T. granarium* in shape and size,. The dorsal surface is moderately clothed with fine hairs. A median ocellus is present between the compound eyes. Number of antennal segments is usually 11 and there is a distinct antennal club consisting of 6-7 segments in males and 4-5 segments in females. The antennae fit into ventral grooves in the prothorax.

Quarantine Measures

- Import of oat grains for consumption from Australia shall accompany PSC with an additional declaration for freedom from variable and with special condition that the consignment is fumigated with methyl bromide @ 80 gm/m^3 for 48 h at 21°C or any other treatment duly approved by the Plant Protection Advisor to the Government of India.

5. Cocoa moth : *Conopomorpha cramerella* (Lepidoptera, Racillaridae)

The adult is a small brown moth and 7 mm in length. It has a wing span of about 12 mm characterized by bright yellow patches at the tips of the forewings. The moths have very long antennae which are swept backwards in their natural resting position. Moths in flight have an appearance similar to that of large, slow-flying mosquitoes.

Quarantine Measures

Import of Rambutan *(Nephelium lappaceum)* fruits for consumption from Thailand shall accompany PSC with an additional declaration for the freedom from *cocoa moth* and with special condition that the consignment is fumigated with mthyl bomide @ 32 g/m^3 for 24 hrs at 21°C and above under NAP. The treatment shall be endorsed in PSC.

6. Tropical nut borer : *Hypothenemus obscurus* (Coleoptera : Scolytidae)

The adult females are 1.2-1.4 mm long and about 2.3 times as long as wide. The colour of the female adult is light immediately after emergence, but darkens within a day to a very dark-brown. The males are approximately half the size of the females and are very rare.

Quarantine Measures

Import of cocoa bean (fermented and dried) for processing or industrial use shall accompany PSC with an additional declaration for freedom from tropical nut borer and with special condition that the consignment is fumigated with methyl bromide @ 16 m^3 for 24 hrs at 21°C and above under NAP. The treatment shall be endorsed in PSC.

7.Coffee berry borer : *Hypothenemus hampei* (Coleoptera : Scolytidae)

Males are often apterous, stunted and deformed. Females with body 1.4-1.6 mm long and 2.3 times as long as wide, entirely black. Antennal funicle usually 5-segmented. Pronotum with fine, raised basal and posterolateral marginal bead, anterior margin bearing 4-8 coarse teeth of about equal size. Elytra with declivity convex, gradual, extending almost to middle of elytra.

Quarantine Measures

- Import of beans (seeds), berries(freshly harvested), grafts, bud wood, seedlingt, rootedcuttings of coffee (Coffea spp.) is prohibited from Africa and South America.
- Beans (seeds), berries (freshly harvested coffee) (*Coffea* spp.) and related species of Rubiaceae can be imported with additional declaration for freedom from coffee berry borer and post entry quarantine for one year period subject to the recommendation, supervision, monitoring and testing by Director, Central Coffee Research Institute, Chikmagalur (Karnataka).

8. Bean bruchid : *Acanthoscelides obtectus* (Coleoptera : Bruchidae)

The adult is 3-4.5 mm long having grey brown and reddish-brown body colours. Antennae are dark grey except for segments 1-5 and 11, which are reddish. The abdomen is reddish. Each hind femur has on the inner ventral ridge near the apex, a large tooth followed by two or three smaller sharp teeth.

Quarantine Measures

- Import of bean grains for consumption shall accompany PSC with an additional declaration for freedom from this pest and with special condition that the consignment is fumigated with methyl bromide @ 32 gm/m^3 for 24 hrs at 28°C.

9. Chocolate moth : *Ephestia elutella* (Lepidoptera : Pyralidae)

The wing span is 16-19 mm. The head and thorax are greyish-brown with the abdomen usually pale. The fore wings are greyish-brown, variably suffused with darker brown. The postmedian and antemedian fascias are dark greyish-brown; the postmedian fascia is double, with a pale line dividing it. The hind wing is white.

Quarantine Measures

- Import of cocoa beans (fermented and dried) for processing or industrial use shall accompany PSC with an additional declaration for freedom from this pest and with special condition that the consignment is fumigated with methyl bromide @ 16 m^3 for 24 hrs at 21°C.
- The plant quarantine officials shall inspect all the consignment of agricultural commodities in India at the port of entry and they shall carry out appropriate inspection methods. In case of interception of insect pests of non-quarantine pests, the consignment shall be fumigated and released. If quarantine pests are intercepted, the consignment shall be fumigated.
- Plant quarantine acts as filter rather as a barrier in the international trade. The stringent Phytosanitary Measures and Regulations are in vogue to prevent the introduction of exotic pests in India. Constant vigil and rapid quarantine screening are the only way to safeguard our agriculture from exotic insect pests.

Chapter 10

Methods of Storage

Since commencement of the settlement of world, the search for food had been uppermost, thereby, practising the conservation of food is a basic need for ensuring its ample supply to human being. The problem of food storage is growing fast day by day with the increase in the population. The existing shortage of protein and caloric requirements will not only be solved by growing more food and resorting to family planning measures but also by saving the produce from deterioration and wastage, and its efficient allocation to meet the growing requirement of population. In countries with varied climatic conditions, it is a steady struggle of man with insects, mites, rodents and microbes in getting the food stuff.

Need and Objectives of Storage

Since times immemorial, man adopted a particular system of storage and certain produce storage standards. Some of the storage methods are still in use in several regions, in the capacity varying from one kg to several hundred quintals, like hollowed gourds, skin bags, mud bins, excavated stones, woven baskets, underground pits etc., particularly in the tropics. The problems of storage in developing countries are more acute than that in well developed countries and a considerable disparity seems to be there in food availability. All need a food supply that should be nutritious, clean; free from insects, rodents, birds, other animal filth and harmful chemical residues. There are two types of needs, one immediately after harvest and the other in off season. There is no need of storing the produce for immediate requirement but for the latter case, the need to conserve produce properly is inevitable. Further, the need for storage has always arisen for one reason to distribute commodity in deficit production areas and the other based on improved quality standards. In all the way, therefore; growers, processors, transporters and warehouse owner must assist to comply with these requirements.

Since harvested produce passes through various channels, storage and transportation losses in quantity and quality are bound to occur between the field and table. The deterioration of produce may commence in the field before harvest due to attack of rodents, insects and moulds which are accentuated, if the produce is not harvested at peak maturity. Rodents and insects travel freely from nearby stores, invade growing crops adjacent to the stores and are particularly active in flight during certain time of the day.

The insect nuisance in tropics has been reported to be serious and related to the type and conditions of storage structures, weather, deprived sanitation and prevailing socio-economic conditions of the farmers. The acquaintance of the farmers to efficient methods of storage can result in more money being available through their talent to sell good quality produce when prices are favourable to them. The financial disparity between the price the farmers obtain for his produce at the time of harvest and that which is paid by the consumer is substantial. The storage method should be scientific which facilitate grading operations because maximum return can only be achieved by fulfilling quality standards.

Storage of food grain has been practised with two main objectives:

1. To meet the needs of mankind in day to day requirement, and
2. To maintain buffer stock to distribute in famine affected and deprivation areas as and when needed.

The previous objective to some degree can be achieved by making the ordinary structures as big as required but the later needs still a design for building of spacious structures. Storage structure is the most important single factor which can keep the losses caused by insects, rodents and fungi to a minimal level. Therefore, the storage design should be such that the spoilage and deterioration can not be permitted from the ravages of all such agencies. The entomological or engineering requirements can not be taken up separately as both are harmonizing to each other. Insect pests and micro-organisms grow with the variation in climate and engineers are to take action, how to eliminate or minimise the climatic effects by developing designs.

Pre-requisites of Storage Structures

Various structures are used in different areas or zones but the one which is more in use should have following pre-requisites:

(a) It should be made up of easily available materials, cheap in construction and safe for storage of grain.

(b) Primarily, the site for structure should have proper drainage.

(c) Fundamentally, an environment within the structure should restrict the supply of oxygen and reduce the temperature fluctuations thus maintaining more or less a steady temperature.

The traditional underground storage practices provide the all proximate conditions. It is interesting to note that the above ground structures in India are made up of materials that have poor thermal conductivity. Efforts have been made by different workers to modify the existing structures of particular zone or locality to make farmers and grain handlers to understand the scientific storage and make use of the information already available to protect the produce. The Indian Standards Instituton, a non-governmental organisation has laid down standards for rural food grain storage structures, *viz.*, *Bukbari*, *Khatti*, bagged food grain storage structure etc. Generally, the standards are adopted after a thorough examination by a team of representatives from the government, trade and research organizations. The standards for storage structures is intended to provide construction which would preclude, to a great extent, the chances of damage to them as well as to the food grain stored within. At the same time it is constructed conveniently with locally available material to suit the people in the rural areas where such structures are prevalent.

Types of Storage

The grains are stored in bulk (loose grain storage) or in bags (sacks), the latter being common in tropics because of ease in handling, more flexible and requires less capital than the bulk storage. The bulk storage has been practised from time immemorial. It is practised for small quantities, whereas, large quantities are stored in bags. The bulk storage can be divided into underground storage and above ground storage.

1. Bulk Storage

a. Underground storage

Khatti or *Khai*: The underground storage system is commonly called *Khai* in Rajasthan and *Khatti* in U. P., Punjab and M. P. This is a method of storage that man started to practise from ancient times. It is a dug out pit with a narrow circular inlet and used in areas where sub-soil water is low. The people in Southern Europe, Middle East and Asia have been using the underground pits to store grain from primeval times. The use of the size and shape and depth of pits vary from place to place, country to country, though the principle remained the same. In India, they are known by different names in local dialect which vary in capacity as per need of the farmer.

Indian standards for this were adopted in 1960. It shall be cylindrical, like a well and waterproof; floor is constructed of the layers of sand (coal cinders, ash from brick kilns or from similar sources), bitumen and cement concrete; the wall shall be of bricks in cement or lime mortar with a coating of bitumen or polythene, and plastered with cement; the top must be of reinforced cement concrete with a central manhole covered with cast iron lid. It shall be located in areas where level of sub-soil water during any time of year is not less than 2 m below the floor; shall be situated in firm, raised, well drained ground at least 6 m from cattle pans etc. and 30 m from brick kilns. This practice was adopted in India by commercial grain handlers and Government farms sometimes five decades back. The dimensions of *khatti* based on different capacities have been revealed in Table 10.1.

Table 10.1: Different capacities and dimensions of *Khatti*

Approx. capacity (tones)	Internal dimensions		Height above ground level (m)
	Diameter (m)	Depth (m)	
5	2.0	1.5	0.75
10	2.0	3.0	0.75
20	2.5	4.0	0.75
40	3.5	4.5	0.75

The main reason of declining the use of *Khattis* is due to the change of climatic conditions, like heavy rain and flood which are responsible for spoilage of grain due to growth of moulds. The rain water percolates through the sides of walls of *Khattis,* thereby making the grain black. This method is not free from moisture ingress. At places where thick layers of straw and dried grasses are in use, the grain is kept safe to a considerable extent. The underground pit has advantage of mostly safe from insect damage because of O_2 depletion and accumulation of CO_2. Pit storage is based on hermetic principles and hence cannot be said to be dangerous to stored grain, however, it needs improvement. Some efforts have been made to improve upon these by making *Pucca Khattis* cemented upground and underground. The *Pucca Khattis* form a reliable storage system that keeps to a minimum losses resulting from insects, moulds and rodents.

b. Above ground storage

Earthen receptacle or Matkabin: Earthen receptacles used by the farmers are of various sizes and shape for keeping small quantities of grain. They are provided with earthen lid to close the vessel when filled with grain. Some cultivators have been using these pots by painting them either with castor oil or bitumen which serves as barrier to moisture ingress and also limits the oxygen supply. It is also evident that in India at many places farmers still use baked earthen vessels plastered with oil based mud paints to preserve grain.

Kanda or *Bhakhar*: A receptacle made up of bamboo sticks or *Saccharum* or *Sorghum halepense* grass sticks plastered with mud and lid at the top is known as *Kanda*. The bigger sized *Kanda* is often termed as *Bhakhar*. The capacity of these receptacles vary from 0.5-15 quintals. They are commonly used for storing grain in villages of Kumaon hills in India. A basket commonly known as *Kihenge* is cylindrical in shape and usually made of interwoven reeds, bamboo splits or thin sticks depending on accessibility of material. The sides of such structures are plastered with a thin layer of cow dung and mud. Similar structures made of bamboo matting or waffle stem of *Saccharum arundinaceum* or sticks of some other plants are found in parts of Bengal, Assam and South India. These structures are in general kept in side room. Wherever they are kept out side, then covered with thatched roof or *Chhapar* and the stores stand on raised platform which may be built with bricks or logs of wood. The capacity of such structures vary from 1.0-2.5 tonnes. The life span varies upto ten years depending on the method of construction and kind of material used. Bins made entirely of straw are in common use in Bengal and some parts of Punjab and *Tarai* in North India like Kashipur.

Thekas or Pallas: In some parts of Punjab, Haryana and Uttar Pradesh (India), a structure known as *Pallas* and *Thekas* are used to store grain. *Papas* or *Palli* is a big rectangular piece of palm leaf or *Sarkanda* matters is rolled round the platform and tied with rope keeping it slightly above the ground. The edges of the mattress are then stitched together. The grain is poured into this drum like structure. When full, it is covered with another piece of mattress. At some places, iron pan is used in place of platform. *Pallas* is of 15-1,000 quintal capacity. *Thekas* or *Thekis* are receptacle made of gunny. They are long in shape and much bigger than the ordinary bag. They are 3-4 m in height and can store upto 150-250 quintals.

A wide variety of material is used to construct them at different places. Cylindrical *Thekas* made from woven reeds of maize stalks and thatched straw roofs are common.

New Hapur Theka : The New Hapur *Theka* (Fig. 10.1), an improved bin already popularised in some rural areas has been designed at the Grain Storage Research and Training Centre, Hapur. It has a round tub-shaped metal base (45 cm high) to avoid moisture ingress from the ground. Eight bamboo sticks (each 5 cm thick) are placed vertically all round in its inner circumference. A cylindrical rubberised cloth is hung inside with the help of hooks fixed at the top of bamboo sticks. After the grain is poured in, the top end of the rubberized cloth is gathered and tied together. The *Theka* is raised 7-8 cm above ground on wooden blocks. The grain is taken out from the top or the sliding door (13 x 13 cm) on metal base. The grain in it can be directly fumigated if needed. The rubberised cloth is found to ward off the rat attack as well. The dimensions based on capacities of New Hapur *Theka* has been revealed in Table 10.2.

Table 10.2: Particulars of dimensions for different capacities of the New Hapur *Theka*

Particulars	Capacity (Quintals)			
	20	17.5	15	12.5
1 Height (cm)	180	180	160	140
2 Diameter (cm)	134	122	122	107

Kothar : A wooden structure made up of planks has been in use in localities where tree planks are easily available. These structures are common in forest areas or hills like Shivalik range of Uttarakhand. A box type structure made up by joining the planks has capacity to store grains upto 20-30 quintals. The joints are cemented with bitumen or mud to make the structure moisture and air proof to some extent. The fabrication or construction of storage receptacles has been done mainly with the material which has been cheaper and easily available in the region.

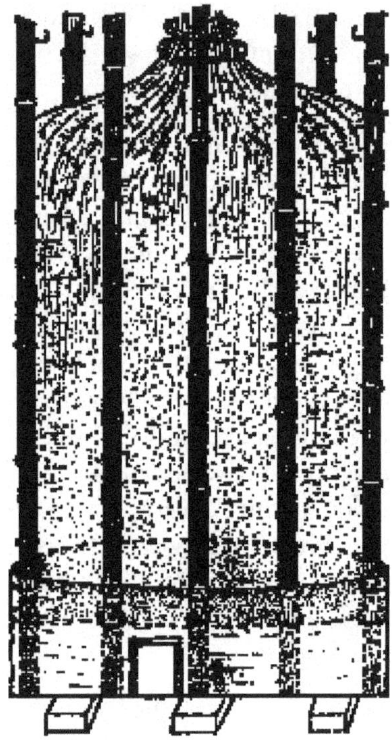

Fig. 10.1 : The New Hapur *Theka* manufactured with bamboo sticks and cylindrical rubberised cloth.

Mudbin (*Kachcha Kothi* and *Kuthla*): The mud bins are known by different names in local dialects such as *Korla* or *Obri* in Rajasthan, *Kuthla* in Uttar Pradesh, *Barola* or *Bharolis* or *Petti* in Punjab and *Khothi* in Bihar.

In Rajasthan, the clay mixed with cow dung and fibres of *Saccharum munja* or wheat *Bhusa* is shaped into a rectangle or round and allowed to dry. The structure is erected on 4-6 pillars (30 cm high). A slab or base is prepared with a thickness of about 10 cm on the pillars. The walls are raised on the slab bit by bit in parts. The walls are 8 cm thick and raised upto a height of 2.5 m depending upon the amount of grain to be stored. An inlet and outlet holes are made at the top and side, respectively. The grain is poured in through the inlet hole; while for taking out the grain, outlet hole is used. In parts of the Punjab, chimney shaped bins are also found. The *Petti* looks like a box which is used in parts of Punjab and is build up of unburnt pond clay. It is either partitioned into different chambers or may not be partitioned. The lid is a big slab made up of clay or wood with holes in each chamber. The holes are plugged with mud and opened whenever grain is to be taken out.

The mud structure is not built all at one time. It is heavy and cannot be made inside the house as it takes long time to become dried and stiff. The base is made with legs and dried in sun again a part of the structure is made and dried; the other parts are built separately and dried in sun. After the parts are dried, they are shifted to a suitable place in the house and all parts are joined together.

Pucca Kothis (*Bukharis*): In order to ease the storage of grain, the living rooms are partitioned. The house may be built of clay or brick and cement work. These are known as *Pucca Kothis* or *Bukharis*. A *Bukhari* may be opened from the top to fill the grain and having outlet hole to take out the grain. These structures may be of different shapes, rectangular, oblong or even circular. In general, 3 m x 5 m size of structure is used. This may be bigger as per the need to store grain. Generally, the capacity of *Bukharis* is about 100-200 quintals. In Bihar, outdoor structures of different shapes are built on a raised platform to avoid moisture ingress. Wheat, sorghum, barley, maize, paddy and peas are stored in such *Bukharis*.

Indian standard for *Bukharis* was adopted in 1955. The standard lays down that it shall be cylindrical, resting on an elevated platform supported on timber posts, or brick or stone masonry pillars; floor made of either bamboos plastered over with mud or timber planks; walls made of cotton sticks, or reeds or bamboo or bamboo splits plastered with mud, and reinforced at intervals with timber fixed firmly throughout the height from ground; roof shall be of conical framework made of cotton sticks or bamboo and thatched over with well dried straw of wheat or paddy etc. and plastered with mud. An opening about 6 inches below top edge of wall for filling the grain and another for taking it out is provided at the floor level, but both are kept sealed with wooden plank when the structure is closed. The site of the structure must be at least 20 feet away from cattle barns and 100 feet away from large trees, stocks of hay etc. which constitute fire hazard. The different dimensions based on capacities of *Bukhari* have been given in Table 10.3.

Table 10.3: Capacities, dimensions and number of pillars required for the construction of *Bukhari*

Approx. capacity (tonnes)	Internal dimensions (ft)		No. of pillars
	Diameter of floor	Height of walls from floor level	
3.6	6	7	4
9	8	9	5
18	10	12	9
36	13	14	13

Pusa bin : The Pusa bin has been designed by Pradhan and Mukherjee (1968) at Indian Agricultural Research Institute, New Delhi to demonstrate the utility of polythene device (Fig.10.2). The bin required for storing 2,000 Kg grain measures about 1.40 m long, 1.00 m broad and 1.60 m high (internal dimensions). The four walls consist of two layers of mud or unburnt bricks (each layer 7.6 cm thick) with a polythene film sandwiched in between them. The floor and the roof comprise of similar structure with each layer 5 cm thick. The roof is supported on a wooden framework. The manhole is 60 x 60 cm in a corner of top surface. About 9 m of polythene or alkathene (700 gauge), 180 cm wide (preferably ISI brand) is required; 750 *Kachcha* bricks (22.9 cm long, 11.5 cm broad and 7.6 cm thick) are used. An outlet pipe is kept 9 cm in diameter and 24 cm long with a tight fitting lid.

Metal drums: Generally, the storage in metal containers enables spoilage through moisture migration particularly in case of grain like paddy which has invariably higher moisture content. Metal drums of various sizes have been used by the farmers in India and other countries. These seem to have greater potential as air tight storage containers.

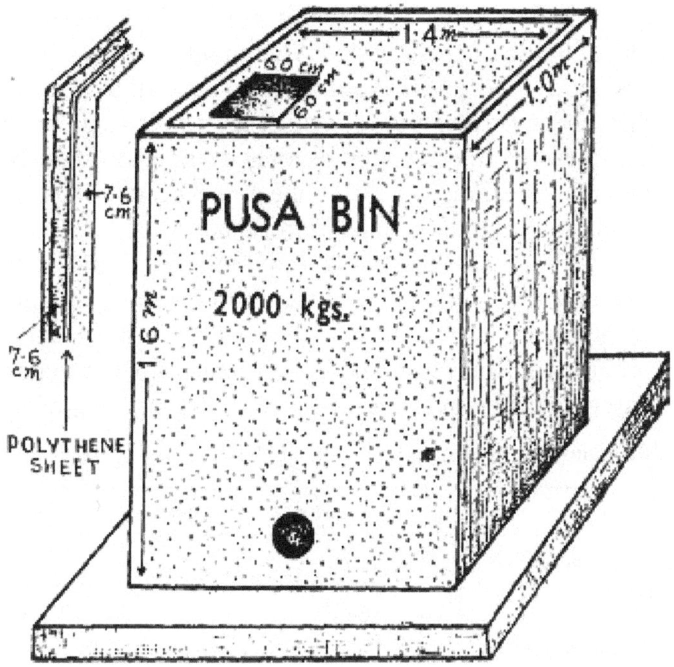

Fig. 10.2 : Pusa bin showing dimensions and a polythene film sandwiched in between two layers of mud walls.

(a) *Mobile oil drum and coal tar drum bins:* The simplest form of metal drums is the open mobile oil drums or coal tar drums which have been extensively used by the farmers. A lid over it is manufactured with the help of local fabricators, therefore, to enable them for effective fumigation.

(b) *Domestic Hapur bin:* The Indian grain Stoarage Institute which is engaged in the development and dissemination of advances in storage technology to users has developed metal bins for domestic storage of food grains. These are made of galvanized iron and/ or aluminum sheets. The bins are available from 200-1000 Kg capacity.

(c) *Aluminium bins:* The Aluminium bins (Fig.10.3) manufactured by the Indian Aluminium Co. Ltd., Calcutta have been tried, and the results of experiments have indicated that wheat, Bengal gram and other food grains may be stored satisfactorily in these. For prolonged storage in areas where the relative humidity is high, the air space at the top should be minimum. Indian Standard for this was adopted in 1961. It is cylindrical with conical roof, composed of aluminium sheets and constructed on a platform or plinth 60 cm high with the spout embedded in the plinth itself; a manhole is provided in the roof. The dimensions of Aluminium bins based on different capacities have been given in Table 10.4.

Table 10.4: Dimensions of aluminium bin of different capacities (tonnes)

	15 tonnes	28 tonnes	35 tonnes
1. Body diameter	370	430	430
2. Overall height	275	350	430
3. Height of cylinder	190	250	325
4. Platform diameter	395	450	450
5. Height of platform	60	60	60

Fig. 10.3 : Prefabricated aluminium bin for grain storage

Chittore stone bin: The locally available materials should wherever possible be used in the construction of Chittore stone bin. In Rajasthan, stone slabs are naturally available and abundant. At the

College of Agricultural Engineering, Udaipur; a stone bin called the Chittore bin has been developed using 40 mm thick stone slabs. It is rectangular bin of 250 kg capacity and is constructed by the farmers themselves using mud as a cementing material.

Elevator storage: Elevators are equipped with devices based on modern concept. The grain elevator at Hapur, obtained under USAID programme and erected by Indian technicians in 1958 has a storage capacity of 400,000 tonnes food grains. It has 20 silos or storage bins (76 feet high and 22 feet diameter) which have a total capacity of 10,000 tonnes. Additionally, the head house (156 feet high and 28 feet in diameter) contains grain elevators, auxiliary tanks, grain cleaning and disinfecting equipment, grain driers and weighing machines. Besides the elevator, there are masonry godowns with storage capacity of 10,500 tonnes. There are rail hoppers on one side of the silos, parallel to the railway line and two truck dumps on the other side. About 24 feet underground is an elevator pit to receive the grain from the rail hoppers and the truck dumps. After the grain is unloaded into the rail hoppers or the truck dumps, it is transferred by conveyor belt to the elevator pit from where it is lifted by the elevators to a height of 136 feet, and after initial weighing passes through the cleaning machines and then on to the automatically integrated weighing process which indicates the quantity fed into each bin. The grain then drops on to another conveyor belt which takes it to the various storage bins.

A drier tank along sides of the head house removes the moisture from the grain by hot air produced from an oil burner. Ten tonnes of grain per hour is handled by this way. Bins are also provided with electric temperature indicators to detect the environment within, and air can be drawn to cool them when unduly heated; grain can be readily fumigated within, if necessary. In order to take out the grain, the conveyor belts at the bottom of bins transfer it to the elevator pit where it is automatically weighed and removed in bulk or bags. Two diesel electric generators of 80 KW each operate the elevators; a third generator serves as a stand by in an emergency.

The elevator can handle about 100 tonnes of grain per hour with the help of one controller and nine operators. Various kinds of grain, *e.g.*, wheat, paddy, gram, pearl millet etc. can be stored simultaneously. Further, grain is safely stored upto five years in an area about one-fifth of that needed for conventional storage godown.

Vacuum process storage: The vacuum process storage (VPS) is a modern concept in storage system. Under the VPS system, the PVC containers manufactured for the purpose are filled with food grains and air is sucked out of them instantly. Thus vacuumised containers could be operated like stone boxes and stacked like gunny bags in the open field. The food grain in such containers never gets infected or catches fungal infection in the absence of the air. Besides this, the food grains thus stored never get discoloured and suffer damage in quality even though the stocks are kept for 3-4 years.

2. Bag Storage

The socio-economic conditions of farmers in tropical regions where the means of transport are poor, the bag storage is practised for large scale storage. The grain is dried to a moisture level of 8-12 per cent before storage. Bag storage requires considerable labour but is undoubtedly the most satisfactory one, if absolute smallest amount is to be spent on permanent storage facilities and equipments. A farmer does not want to spend big amount of capital at one time in constructing a big storage structure. Bag storage has merit of short term storage, therefore, before going for a type of storage, it is always better to decide how long a grain lot is to be stored. Short term storage can be done under the roof of corrugated iron, a plastic covering, where grain is intended for very early onward movement. Storage period has a very big effect on storage type. For short term storage, no control measures would be needed, and for seasonal storage in tropics where it is a general practice, some insect control treatment needs to be applied. It is obvious that seasonal storage usually bear the maximum cost.

In long term storage, insect control is warranted in any climate. It requires more care than short and seasonal storage and consequently, it is appropriate to be regarded as necessarily more expensive. When once the length of storage is decided, this is very essential for tropical regions than temperate regions. It needs pre-planning, where the method and type of storage can be appropriately planned. Another, most important point to be accounted is the condition of the grain and the environment in which the grain is to be stored. At certain times due to unfavourable unforeseen rains, grain is harvested at high moisture content or absorbs moisture after being harvested. In tropics, such occasion faced is seldom but in temperate region grain is generally stored at a high moisture levels. Whenever grain is wet due to rain and cannot be dried in sun or with artificial drying means, 0.5 per cent glacial acetic acid or vinegar may be sprinkled at the rate of 1 litre of solution on one quintal of grain. This would keep the grain under check from fungus invasion and deterioration due to biochemical changes. This grain may later be dried and stored properly.

Hessian bag (Jute bag) and plastic bag storage came into existence when difficulty in transportation of grain was felt. Cultivators felt it easy to pack grain in cloth and move it from one place to another. Bagged grain is more convenient in handling than grain in bulk, though it requires more labour. Gunnies, easily absorb moisture and encourage insect entry and retain it for quite sometime even after fumigation. Bagged grain stored in damp area soon comes to equilibrium with atmospheric water vapours. Rats also have an easy access to bags. Both initial and maintenance cost of bags is heavy. It is inevitable with the conditions inscribed that bagged grain once infested and left uncared, may become unfit for human consumption or milling. Therefore, the safety of the grain stored for long period is dependant upon the place of store and elimination of insect infestation.

Bagged Food Grain Structures and Their Maintenance

Most of the structures have been used to store grain to be used for household consumption or seed to sow it in the next season crop. But with the increase in population and high production of crops made the man to think of safeguarding the produce. The campaigns to increase in production of grain, like Green Revolution was launched with the main objective to guarantee a year round availability of food at every reasonable price at one hand and create reserve food on the other. Some of the grains are stored in bags because of limitation of bulk storage. Wheat and paddy stored by farmers for food and seed in bulk needs extra care, may be in vertical or horizontal system of bulk storage with facilities of aeration. In bagged grain storage protection against insect invites special attention and importance of good warehouse hygiene. Any portion of a store where bagged consignment is received should thoroughly be swept and the sweepings are burnt. The walls should be brushed down and whenever possible, sprayed with malathion or other suitable insecticide immediately prior to sacking and subsequently at regular intervals.

The maintenance of stores is dependent upon the structure built. The warehouses hygiene is maintained and effective pest control is carried out where warehouse is made of concrete or cement block construction with smooth interior finish without cavities and projections. The ceiling should be contiguous with the walls, and constructed in such a way as to ensure airtightness and ventilation between ceiling and roof. The ventilation of the storage space should be kept to a minimum and should be such that complete ventilation control is possible, allowing it to take place only during period of low ambient relative humidity. The ventilator is fitted with fine mesh screening net for excluding flying insects. The floor should be kept 4-5 feet above ground level and incorporate bitumen to check water vapour, bitumen impregnated hessian cloth (Tarfelt) should be used in roofing. There should be adequate concrete drainage with concrete splash aprons. The bags should be stacked upon wooden plateform and provide proper drainage to prevent the uptake of moisture from the floor. The drainage system should be made up of Tun, Sal or other wood that is specifically recommended for making pallets. It should be well painted. The drainage be always cleaned very carefully. The bags should be stacked in an uniform size to facilitate fumigation under fumigation covers. The density of bag is another very important point to be checked to prevent from (i) tearing off of the jute bags, and (ii) the loss of viability due to pressure on the bottom most bags. Stores are normally constructed for 500 to 1000 tonnes capacity. The average density of 846 kg/ m^3 for bagged millet in stack has been recommended. The bulk density, further, indicated was from 800 kg/m^3 at 9.5 per cent moisture and 750 kg/m^3 at 15 per cent moisture. A bag density for normal storage at a normal moisture level should be 846 kg/m^3 which agreed with the earlier recommen-dation of 830 kg/m^3 for bagged millet.

According to Hayward (1981), the average bag density should not be more than 846 kg/m^3 and height of the layer of bags in a stack be about 13-15 only in case of pearl millet. The bags made of plastic film like polyethylene are recommended for storage of produce. They are made of fairly heavy thickness (400 to 700 gauge or 100 to 200 microns in thickness) film which also serves the purpose of air tight storage. There are several species of insects, like *Rhyzopertha dominica* (Fab.) and Bruchids which puncture the film. It is, therefore, suggested that inner lining of polyethylene may be given with fine cloth to prevent penetration of insects. In order to protect grain from being spoiled, chemical treatment of bags becomes necessary.

Commercial stocks, the actual marketable surplus comprise 30-35 per cent of the total production. Probably, above ground *Kothis* can best protect these large stocks, particularly wheat without much damage and whatever damage on account of heat is restricted to periphery, though infestation intensifies when grain is disturbed. Prevalently, commercial stockists including Warehousing Corporation store grains in bags which are stacked in warehouses called godowns, for which ISI has standardised the structures. Bag storage is now called conventional storage, and recent specific conventional storage designed can store about a ton in 6 square feet area safely against rodents, birds, termites etc under optimum ventilation, light and damp proof conditions. Of course, bulk storage provides better protection than bag storage because the heat developed by insect metabolic activity within the bag of grain is not built up to the extent detrimental to insect activity and is dissipated through inter-bag space. But free fatty acids increase relatively fast and gluten content decreases more rapidly in bulk grain consequent to any deterioration therein.

Indian Standards for Construction of Bagged Food Grain Storage Structures

Indian standard for this was adopted in 1965. Such structure should be located on a raised well drained site not liable to flooding or inundation. It shall be 0.5 Km away from kilns, dumping grounds or such sites, the vicinity of which is deleterious to the safe storage of grains. Further, it must be 30 m away from dairies, poultry farms and 150 m from factories or other possible sources of fire, *e.g.*, fire workshop, petrol pump, hay stacks etc; also that its construction in residential area be avoided as far as possible. There shall be no tree nearby because its roots might affect the foundation; precautions should be taken against rats, squirrels, termites etc. It may be preferably located near a transport junction or a main road. The capacities and dimensions of bag storage structures have been given in Table 10.5.

Table 10.5: Capacities and dimensions of bag storage structures

Approx. capacity		Internal dimensions	
Bags	Tonnes	Length (m)	Breadth (m)
1050	100	12	7.5
2625	250	20	9
5250	500	27	12
10500	1000	33.5	18
21000	2000	50	27

Stacking : The floor shall be marked into rectangular or square stack spaces not exceeding 30 x 20 ft, keeping in view that the alley ways of 3 ft wide are left in between stacks and the walls, and also in between the stacks. The dunnage comprising of timber squares (4 x 4 '') or timber pallets (rectangular platform, 8 x 5 feet with 1 foot space in between transverse battens) are laid, and the floor space in between them as well as the dunnage itself shall be dusted with malathion dust 5 per cent. The stack shall be built over the dunnage in cuboid shape and again dusted with malathion dust 5 per cent. The maximum height (in bags) shall not exceed the limit set for the different food grains as given in Table 10.6.

Table 10.6: Number of bags in stacking of different commodities.

	Commodity	Maximum no.of bags
(i)	Wheat, maize, barley, paddy, oats, gram, unmilled pulses, sorghum, millets	16
(ii)	Rice ad milled pulses	14
(iii)	Wheat flour, gram flour and other milled products	12

For fumigation, the stack is covered with a loose fitted, cubiclal gas-proof cover stitched out of 3 ply rubberized cloth or other suitable gas proof fabric and provided with a few openings (about 6 inches in diameter) carrying a leather tube of 9 inches long, on the top face. Lower edges of the cover are folded inwards and plastered on to the floor with mud (not containing sand, stones, dry earth etc). EDCT mixture at the rate of 30 lbs per 1,000 ft^3 space within the cover is poured at the top of the stack through the tubes which are tied soon after and the entire cover kept closed for 48-72 hours; gradually now, the edges of the cover are lifted and left open for six hours after which the entire cover may be removed.

Loose Storage

Bag Storage

Bag Storage in Room

Bhukhari

Stone bin

Earthen pot

Fig. 5: Different methods of food grain storage in Rajasthan

Hanumangarh

Bundi

Ajmer

Sirohi

Jhunjhunu

Sikar

Fig. 6: Different types of metal containers used in Rajasthan

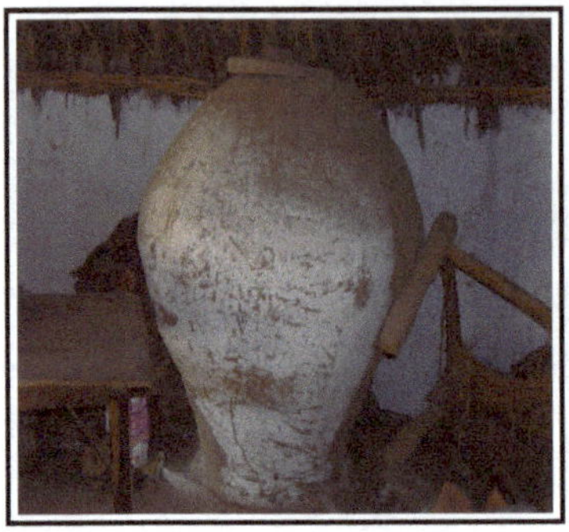

Barmer

Tonk

Churu

Jalore

Sikar

Sirohi

Fig. 7: Different types of Kothies used in Rajasthan

Chapter 11

Food Grain Storage Organizations in India

A series of efforts have been made to initiate the work on post harvest grain technology in India. The history of post harvest grain technology in India goes back as early as 1888 when COATS published several papers from Calcutta related to biology of stored grain pests. The first professional was appointed in the year 1905, known as Imperial Entomologist, Harold Maxwell Lefroy adorned with this post. Lefroy (1906) for the first time gave a tentative list of storage insects, which was further supplemented by Fletcher (1914). The problems of grain storage were discussed at the entomological meetings held at Pusa, Bihar in the years 1917, 1919 and 1923. The first step towards establishing an Institute in India was in the year 1950 when Central Food Technology Research Institute (CFTRI) was established at Mysore (Karnataka) with a mandate to train personnel and evolve technology related to food processing and preservation. The Central Warehousing Corporation was set up on 2nd March 1957 under the Agricultural Produce (Warehousing and Development) Act 1956. The Food Corporation of India was established on 1st January 1965.

In India, the farmers in traditional bulk and bag storage for food, feed and seed stored about 65-70 per cent of the total food grains and seed purposes till next harvest. Rest 30-35 per cent are procured and stored by food grain organizations, like Food Corporation of India (FCI), Central Warehousing Corporation (CWC), State Warehousing Corporations (SWCs), cooperatives, rural godowns and other state government agencies. In 1958, the Department of Food established a centre at Hapur known as Grain Storage Research and Training Centre (GSRTC) with the main objective to train the personnel of state government engaged in food procurement, storage and distribution. This centre was upgraded during 1968-1973 with the technical and financial assistance of UNDP and GSRTC and redesigned as Indian Grain Storage Institute (IGSI). The field stations at Ludhiana and Bapatla (later shifted to Hyderabad in March, 1979) were established in 1968-69. Subsequently, three stations were also established at Jabalpur (1980), Udaipur (1981) and Jorhat (1981).

Food Grain Storage Organizations in the Country

- Food Corporation of India (FCI)
- Warehousing in India
  - a. Central Warehousing Corporation (CWC)
  - b. State Warehousing Corporation (SWC)
- National Cooperative Development Corporation (NCDC)
- Rural godowns
- State governments and their agencies

1. Food Corporation of India (FCI)

History

The executive wing of the Food Department headed by the Director General of Food is responsible for the purchase, storage, transport and distribution of food grains ever since inception of the Department of Food in 1942. A committee appointed by the government recommended the setting up of machinery similar to a business corporation with essential freedom and elasticity of operation. The suggestions were implemented in 1964 with enactment of the Food Corporation Act (no. 37 of 1964). The FCI, thus, came into being as an autonomous body under the administrative control of the ministry and started functioning from 1st January, 1965 with the following food policies of the Department of Food

- Distribution of food grains throughout the country at reasonable prices.
- Effective price support operations to safeguard farmers' interests.
- To manage satisfactory level of operation and buffer stocks of food grains to ensure national food security.

Objectives

- To ensure that the primary producer obtained the minimum price that may be announced from time to time and to protect the consumer from the vagaries of speculative trade.
- To engage primarily in the purchase, storage, movement, distribution and sale of food grains and other food stuffs.
- To undertake state trading in food grains in a commercial manner within the framework of government policy so that these operations would be in the interest of increased agricultural production as well as the common consumer.
- To ensure a strategic and commanding position in the food grain trade in the country and to build adequate buffer stocks.

Functions and Activities

- Purchasing, storing, moving (transporting), distributing and selling food grains and other foodstuffs.
- Building and operating national buffer stock of food grains.
- Setting up or assisting in setting up modern rice mills and other undertakings for processing food grains and other foodstuffs.
- Acting as a price support organization for food grains, mainly wheat and paddy.

- Importing food grains as per policy of the Government of India from time to time.
- Promoting by such means as it think fit, the production of food grains and other foodstuffs.

Producers' and Consumers' Interests

Farmers are offered both market and price security. For increase in food grains production farmer's interests are well looked after. With the presence of FCI and other agencies no distress sale occurs during the marketing season. FCI also reaches to the door steps of farmers for buying the grains through specially operated centers in rural areas. The minimum support prices are assured to the farmers resulting in their prosperity.

Subsidy is provided by the Government of India for the consumers, particularly, those belonging to the weaker sections. Uninterrupted supplies of food grains at subsidized prices are provided round the year. FCI is also ready to rush supplies within 24 hours in any part of the country affected by natural calamity caused by flood, drought or cyclone. FCI's vast network of food storage depots is spread throughout the country for feeding the public distribution system to maintain the supply line on war footing.

National Food Security

The essential requisites for building up of an effective national food security are adequate operational buffer stocks, optimum dispersal of stocks in different parts of the country and an efficient delivery system including fair-price shops and attractive prices. Buffer stocks of food grains are maintained to tide over many unforeseen crisis on the food front. The food management in the drought areas bears ample testimony to the importance of buffer stocks.

Problems and Issues

(a) *Buffer stocks of food grains* : The most important challenge is to build a large reserve of buffer stocks for ensuring the food security in the country. This would ensure not only the easy availability of food grains to the vulnerable sections of the society but also guard against speculative activities of the private trade and wide fluctuations in the open market prices of food grains. The Department of Food constantly reviews the question of the size of buffer stock. According to the existing buffer stocking policy of the government, FCI is required to maintain a buffer stock of 10 million tonnes besides operational stocks. For such a vast country, to estimate the accurate size of the buffer and operational stock in the different months is a cumbersome process.

(b) *Procurement efforts*: The FCI frequently gears up its machinery not only for maximizing the procurement but also for adopting improved and better methods of procurement so that the maximum results are achieved. This means establishment of a number of procurement centers, development of market yards mandis, provision of adequate man and material machinery, so that the grain can be procured under a set of prescribed specifications. Provision of facilities for weighing, marketing, bagging, transportation etc. are pre-requisites and are very complex one. Failure of one or the other will hamper the entire chain of procurement.

(c) *Larger and quicker movements* : This is very essential all over the country for the public distribution system and for maintaining the buffer stocks at various places. The maximum procurement of wheat and rice is being carried out mainly from Punjab, Haryana and Western U.P., while it is to be moved to the farthest end of the country for running the public distribution system efficiently and effectively. It would be necessary to adopt further improvements in this regard with particular reference to movement of grains in bulk.

(d) *Scientific storage :* For holding the buffer stock, scientific storage is a perquisite. The essential aspects are the location of the stocks, logistics of movement, optimum distribution of buffer stocks, development of integrated plans for locating the godowns in alignment with the rail or road movement capabilities etc. Periodic inspection of the stocks, categorization, liquidation of the stocks of low category and damaged stocks on priority basis and maintenance with regular curative and prophylactic treatment besides other hygienic factors are some essential components of the scientific storage of food grains.

(e) *Improvement of preservation technology* : Technology of preservation and storage, specifications or standards of the quality are constantly reviewed for upkeep of the health of stored grain.

(f) *Optimum utilization of bulk storage structure* : The requirement of storage capacity for bag, bulk and CAP storage is constantly reviewed for effective and optimum utilization of the available capacity.

(g) *Organizational and administrative challenges* : On the organizational and administrative side, several emerging problems would have to be met in the future. These problems fall in the following areas

- Streamlining of methods and system in the various operations for ensuring better efficiency and effectiveness.
- Adoption of better management practices for improving the management of the operations.
- Intensification of the measures already initiated for the reduction in wastages, deterioration and losses both in transit and storage.
- Improvement in the management information systems.
- Improvement in the monitoring procedures for quicker decision making.

2. Warehousing in India

Warehousing may be defined as the assumption of responsibility for the storage of goods. It is new in the Indian economy. Warehouse is the scientific storage structure especially constructed for the protection of the quantity and quality of stored products. It may be called the protector of national wealth. The produce stored in the warehouses is preserved and protected against rodents, insect pests, ill effects of moisture and dampness. Warehousing in India is an integrated scheme of scientific storage, rural credit, price stabilization and market intelligence and is intended to supplement the efforts of cooperative institution. In 1928, the Royal Commission on Agriculture felt the need of warehousing system in India. The All India Rural Survey Committee of the RBI (set up in 1951 and submitted its report in 1954) also made comprehensive recommendations for the development of warehousing as an integrated scheme of rural credit and marketing. The Government of India enacted the Agricultural Produce (Development and Warehousing) Act 1956. The act defines the specific functions and the area of operations of Central and State Warehousing Corporations. It encompasses the list of commodities meant for storage. The act provided for (a) the establishment of a National Cooperative Development and Warehousing Board, (b) the establishment of the Central Warehousing Corporation (subsequently established in Delhi on 2nd March, 1957), and (c) the establishment of State Warehousing Corporations in all states in the country (established in various states between July, 1957 to August, 1958).

Objectives

The existing warehousing programme was started in 1957 with the main objective of providing scientific storage. The details of objectives are as follows:

- To improve the standard of living.
- To give better price for farmers' produce at the market.

- To give credit at the concessional rate of interest and loan to the extent of 75 per cent of value of his produce at the time of warehousing.
- To provide better and standard scales and weight which are manipulated against the producer and seller.
- To protect him from other kind of reduction as religious and other charitable purposes.
- To safeguard farmers' and producers' interest against malpractices of the existing marketing.
- To eliminate storage loss caused by the insects, mites, rodents, moisture, birds and squirrels.
- To help in solving the food problem.
- To store improved type of seeds, manures, fertilizers and other perishable commodities in specialized storage.
- Warehouses help in growing more food.
- To give new gunny bags to the depositors at reasonable storage rent.
- To provide better and important link in the chain of agricultural marketing.

a. Central Warehousing Corporation (CWC)

This corporation was established as a statutory body in New Delhi on 2nd March 1957. Under the new act, the Central Warehousing Corporation was formally re-established on March 18, 1963. The corporation made a modest start with 7 warehouses, each of 7000 tonnes capacity. In December 1957, it has set up 330 warehouses in different places with a total storage capacity of 3.789 million tonnes at the end of 1981. The areas of operation of the CWC include centers of all India and that of inter state importance. The CWC is expected to provide another 1.9 million tonnes of warehousing capacity by the end of sixth five-year plan. The functions of CWC are as follows:

- To acquire and build up godowns and warehouses at suitable places in India.
- To run warehouses for storage of Agriculture produce, seeds, fertilizers and notified commodities for individuals, cooperatives and other institutions.
- To act as an agent of the government for the purchase, sale, storage and distribution of the above commodities.
- To arrange facilities for the transport of above commodities.
- To subscribe to the share capital of State Warehousing Corporation.
- To carry out such other functions as may be prescribed under the act.

Besides the conventional storage godowns, the CWC is running air conditioned godowns at Kolkatta, Mumbai and Delhi and also provides cold storage facilities at Hyderabad.

b. State Warehousing Corporation (SWC)

Separate warehousing corporations were also set up in different states of the Indian Union. The first warehouse was set up in Bihar in 1956. At the end of March, 1981, State Warehousing Corporations were operating 1,850 warehouses with a total capacity of over 50 lac tonnes. The areas of operation of State Warehousing Corporations are centres of district importance. The concerned state government and the CWC contribute the total share capital of SWC equally. The SWC's are under dual control of the state government and the CWC.

Working of Warehouses

The warehouses work under he respective warehousing acts passed by the central or state governments. They are licensed under provisions of the act.

Eligibility : Any person may store notified commodities in a warehouse on agreeing to pay the specified charges. The person is required to bring his produce to the warehouses for storage.

Warehouse receipt (Warrant) : This is a receipt or warrant issued by the warehouse manager or owner to the person storing his produce with them. This receipt mentions the name and location of the warehouse, the date of issue and description of the commodities including the grade, weight and approximate value of the produce based on the present price.

Use of chemicals : The produce accepted at the warehouse is preserved scientifically and protected against rodents, insect pests and other infections.

Financing : The warehouse receipt serves as a collateral security for the purpose of getting credit. The commercial banks advance up to 75 per cent of the value of produce stored in the warehouses.

Delivery of produce : The warehouse receipt has to be surrendered to the warehouse owner before withdrawal of the goods. The holder may take delivery of a part of the total produce stored after paying the storage charges.

License for Running Warehouse

The main provisions of the act governing the grant of a license to warehouse are as follows:

- Any person including a company, association or corporate body may apply to the state government for the grant of a license to carry on the business of warehousing.
- The government grants the license after examining the warehouse building and the financial soundness of the party after realization of the prescribed fees.
- The license has to be renewed periodically on payment of prescribed fees.
- The warehouse owner is authorized to receive only notified commodities for storage in his warehouse and issue receipt in a prescribed form.

- It is responsibility of the warehouse owner to keep the premises clean, keep different lots of goods separately in the warehouse and carry on such operations as and when necessary to protect the goods against losses from damage.

Number and Capacity of Warehouses

The number and capacity of CWC and SWC has increased over time. The total capacity of warehouses increased from 3.57 lac tonnes in 1960-61 to 93.16 lac tonnes in 1981-82. It has been proposed to increase the storage capacity of CWC and SWC by 41.20 lac tonnes during the sixth five year plan period (1980-85). The statewise availability of CWC and SWC warehouses at the end of December 1978 in the country is increasing regularly.

Types of Warehouses

(i) On the Basis of Ownership

(a) *Private warehouses*: These are owned by individuals, large business houses or wholesalers for the storage of their own stocks. They also store the products of others.

(b) *Public warehouses*: These are the warehouses which are owned by the government and are meant for the storage of goods of any member of the public against a prescribed storage charge.

(c) *Bonded warehouses*: These warehouses are specially constructed at a seaport or an airport and accept imported goods for storage till the payment of customs by the importer of goods. These warehouses are licensed by the government for this purpose.

(ii) On the Basis of Type of Commodities Stored

(a) *General warehouses*: These are ordinary warehouses used for storage of most of food grains, fertilizer etc. No specific requirement of the commodity is there regarding the especially in the construction of warehouses.

(b) *Special commodity warehouses*: These are warehouses which are specially constructed for the storage of specific commodities, like cotton, tobacco, wool and petroleum products.

(c) *Refrigerated warehouses*: These are warehouses in which temperature is maintained as per requirements and are meant for such perishable commodities as vegetables, fruits, fish, eggs and meat.

Main Reasons for Poor Utilization of Warehouses

- Lack of knowledge about the facility of warehousing available for the farmers.
- Location related disadvantage of warehouses to most of the cultivators located in villages.
- Complicated and time consuming procedure of depositing and withdrawing the produce from warehouses.
- Non existence of nationalized banks in the villages and the problem of arranging the finance at the time of taking delivery of the receipt from the bank.
- Small quantity of surplus produce available with most farmers and the pressing need for finance.

Causes of Slow Progress

The following are the main causes of slow progress of warehousing in the agriculture sector of India:

- Indian farmers are small land owners; obviously, the marketed surplus available with them is small; often, it is not worth while for them to store the produce in a warehouse.
- Indian agriculture is largely dependent on the monsoon.
- Agriculture products are more perishable than industrial products.

- The transport facility from the villages to the cities is not easily available.
- The cost of warehousing per unit of the produce is high.
- Warehouse receipts are papers having no intense value, unless the lenders are sure that these receipts are backed by tangible assets. Often, the lenders are not interested in lending this collateral security.

Suggestions for Improvement

- A network of rural storage centres should be built on a priority basis in order to prevent distress sales, wastage and loss arising out of inadequate and defective storage facilities.
- These storage centres may be constructed and managed by *Panchayats*, co-operatives and other suitable agencies selected by the state government.
- These centres may have a storage capacity of 100- 250 tonnes, mainly for food grains and other agricultural commodities.
- The cost of construction of these structures may be met by 50 per cent bank loans.
- Each rural storage centres should have a manager, preferably from the area served by the centre. The manager should be trained in the basic essentials of warehousing by attaching to a warehouse of the CWC or SWC.

3. National Cooperative Development Corporation (NCDC)

This was set up with the Agricultural Produce Corporations Act, 1956 on 1st January, 1956; later replaced by the National Cooperative Development Corporation Act, 1962.

4. Rural Godowns

These godowns are small sized, ranging from 100-500 tonnes capacity except few godowns which may be as large as up to 3000 tonnes. These godowns are located essentially in rural areas. Since, this scheme was taken up in late 70's and the godowns are small sized, it has been possible to build up a capacity of about 1.5 million tonnes only over the period of last 10 years. Subsidy is provided to the tune of 5 per cent by the central or state government.

5. State Governments and their Agencies

Various state governments and their agencies now gradually build up storage capacity to meet the requirement of public distribution system in order to supplement the efforts of FCI, CWC, SWC's and other agencies. According to the latter reports, a capacity of about 70 million tonnes is reported to be available with various state government and their agencies for food grain storage.

Chapter 12

Fumigants : Properties and Residues in Stored Agro-Products

Introduction

The increase in agricultural production, particularly food grain is one of the major problems of national planners to achieve the target in the food production. Our farmers have been utilized more and more the modern methods of agriculture including the use of chemicals for protection of crop and stored products against the pests. A successful farmer knows that what pests to expect on his crops and stored products, when and under what conditions their attack will be possible. But due to insufficient knowledge of chemicals and their application, our farmers use them without care and injudicious manner, therefore, these chemicals not only give poor kill of insects but also cause the hazards to man and domestic animals. Success in pest control operations depends on many factors such as correct identification of pest, proper timing of insecticidal application, right choice of chemicals and their formulations, doses and method of application.

Fumigants are still widely used for the control of insects and other pest organisms. Because of their unique characteristics and the great adaptability of fumigation technique, fumigants can often provide effective and economical control where other forms of pest control are not feasible. The development of light weight plastic sheets to enclose spaces or materials requiring fumigation has extended its use and made control procedures easier and much more adaptable. Among the fumigants, aluminum phosphide, ethylene dibromide (EDB), Ethylene dibromide- carbon tetrachloride (EDCT) etc. are most common in tropical and sub-tropical while methyl bromide is prevalent in temperate regions. Modern technology and research have also brought to notice certain problems with fumigants that are previously unknown. Among these, adverse effects on seed quality are most important. Thus, fumigants used in seed stores should not affect the viability of the seeds, should not be persistent, non-corrosive and highly diffusible.

History

The concept of the fumigation started 1000 years back with the use of Sulphur fumes for disinfestations of homes. The great development of fumigants for insecticidal purposes took place during the past 150 years with the introduction of CS_2 in 1854 and the use as fumigant in 1879 for the control of stored grain pests. The use of hydrocyanic acid gas (HCN) as a fumigant against cottony cushion scale, *Icerya purchasi* was done in 1886. Chloropicrin was first proposed as a fumigant in 1907, ethylene dioxide in 1927 and methyl bromide in 1932. EDCT mixture was extensively used in India from 1948, then phostoxin or aluminum phosphide and EDB were used after 1948.

Fumigants

Fumigants are chemical substances either liquid or solid; volatile at normal room temperature, convert to poisonous gases, used to kill insects, nematodes, rodents etc., and its application is limited to plants or agro-products in tight enclosure or to soil. Process of exposing pest or pest-affected stocks to the fume of chemical at a lethal dose in an airtight space is called fumigation.

Properties of Fumigants

They have simple chemical formulae compared with many other pesticides and possess the following properties:

(a) *Physical properties*: Fumigants are able to exist in the gaseous state at ambient or other temperature and pressure required. They also diffuse very readily in air and through granular commodities and materials such as soil.

(b) *Biological properties* : All fumigants act as respiratory poisons and are potentially lethal to all stages (eggs, larvae, pupae and adults) of pest organism, provided that adequate concentration of gas are achieved for a long enough period.

(c) *Environment properties*: Fumigants are generally non persistent and are present in treated materials for a relatively short period of time. However, treatment at high dosage or repeated treatments can cause the accumulation of significant residues. It happens when the fumigant reacts chemically with the constituents of the treated products.

Mode of Action

The poisonous gas enters into the larvae, pupae and adult insects through the spiracles and into the eggs through chorion. The toxic action of the fumigants to insects and warm blooded animals results from its ability to combine with iron atom present in iron containing respiratory enzymes such as cytochrome oxidase, succinic dehydrogenase resulting in the mortality of the insects/animals.

Characteristics of An Ideal Fumigant

An ideal fumigant should have following characters:

- Volatile with deep penetration power.
- Extremely toxic to all stages of life.
- Able to distribute uniformly without settling down.
- No adverse reaction with commodities, such as effect on nutritive, cooking, biochemical and other properties of grains and its products.
- Less toxic to plants and mammals.

- Least absorption and high diffusion power.
- Should have distinct warning odour.
- Easy to handle and transportation.
- Should not leave harmful residue.
- Non-explosive.
- Should not be corrosive to equipments and household articles.
- No effect on viability of seeds.
- No consumer hazards.
- Easily and cheaply available.
- Known antidotes and treatments.
- Least soluble in water.
- Should be chemically stable during storage.

Application of Fumigants

a. Fumigant can be applied

- To enclosed areas like mills, factories, godowns, hotels, ships, aeroplane.
- To soil for killing soil insects like wireworms, termites, grubs and nematodes.
- To plant, nurseries and green houses.
- For quarantine measures for import or export.
- To intestinal tracts of animals for killing intestinal worms.

b. Methods of fumigation

(a) *Space fumigation*: The entire space in the godowns is considered for treatment. This method has many advantages such as it helps in destruction of all insects hiding in the cracks and crevices, besides those infesting bagged seeds.

However, in space fumigation quantity of fumigant is used as per the volume of store irrespective of its occupancy. The success of fumigation largely depends on making room sufficiently air tight for particular length of time. For narrow cracks and small holes, stick masking tape or paste craft paper. Plaster of Paris can also be used for sealing of small holes. Similarly, large cracks should be repaired or sealed with suitable sealing material. Outside opening like doors, windows or ventilators should be covered with high-density (1000 gauge) polythene sheet and joint should be plastered with suitable sealing material or adhesive tape.

(b) *Tent or cover fumigation*: Sheet or cover fumigation is commonly practiced in sheds. It provides better gas tight condition and control of insects than space fumigation. The stacks of bagged seeds are covered with gas proof sheet made up of high-density polyethylene sheet of 1000 gauge thickness with perfect sealing at joints or rubber sheet having provision for dispensing phosphine tablets. The most important points to remember are that the size of cover should be sufficient enough to cover stack properly with margins at the ground level for proper sealing. The dose should be worked out according to the quantity of seed than size of cover. It should also be protected from rodents which damage seal sometimes causing loss of gas.

(c) *Vacuum fumigation*: The primary object of vacuum fumigation is to hasten and improve the penetration of the fumigant into the material undergoing treatment. It was originally developed for this purpose when hydrogen cyanide (HCN) was the principal fumigant. In vacuum fumigation, most of the air in the chamber is removed before the fumigant is introduced. It is, therefore, necessary to have a specially designed chamber, usually made of steel which is capable of withstanding external pressure of atmosphere. The installation also includes a vacuum pump able to evacuate the chamber in not more than 10 to 15 minutes, and valves and pipe lines for introduction and exhaust of the fumigant. In vacuum fumigation method:

(i) More rapid penetration of commodities by the gas can be obtained than in atmospheric fumigation.

(ii) The length of exposure can significantly be reduced.

(iii) The removal of the fumigants from the treated area can be speeded up.

(iv) The danger from vapour of fumigants during unloading to worker or technician can be lessened.

In this way, this method is advantageous than the normal fumigation method. Before going for fumigation, it is very essential to consider following points:

- Determine whether or not fumigation is the best way to solve the problem.
- Chose a fumigant that is labeled both for the commodity and structure to be fumigated and the fumigation type used.
- Determine how you will introduce the fumigant and measure its concentration and devise a plan for aerating the fumigated area.
- Much of the information is needed to make these decisions.

Methods of Vacuum Fumigation

In recent years, considerable research has been undertaken on the methods of conducting vacuum fumigation for obtaining the best results with certain fumigants used on certain materials. Basically, there are two main methods: A) Sustained-vacuum fumigation and B). Nearly complete restoration of the pressure, either simultaneously with the introduction of the fumigant or some time later. The choice of one of these methods depends to some extent on the material being treated. Thus, fruit, vegetables and growing plants are usually completely ruined if they remain exposed for more than a few minutes to a pressure below 250 mm (10 in.) of mercury. Seeds, grains, cereals and dried plant products are generally able to withstand these low pressures without any ill effects.

Sustained Vacuum Fumigation

The pressure in the loaded chamber is reduced to between 25 to 150 mm of mercury. The fumigant is then introduced, causing usually only a small rise in pressure. No further alteration is made to the pressure in the system until end of the treatment, which may last for 1.5 to 4 hours, at which allowing air to enter restores time atmospheric pressure.

The fumigant or air mixture is then pumped out. The cycle of air introduction and evacuation may be repeated several times. A process referred to as 'air-washing', until it is considered safely open the door for unloading. This method is used widely all over the world for treating tobacco, grains, flour and meals. Even in this type of vacuum treatment, gas distribution may not be entirely uniform; some chambers are equipped with re-circulating systems and others with fans. Some living plants are able to tolerate sustained vacuum treatments at pressures in the region of 380 mm (15 inch) of mercury. Such treatments may be useful in the control of borers or other insects inside stems or other plant tissues. A substantial number of dormant, non-foliated plants, roots and bulbs can be fumigated safely under a sustained vacuum of 100 mm (4 inch) at an exposure period of 2 to 3.5 hours. In some cases, a higher vacuum can be used without injury.

Steps for Fumigation

1. Preparation
   - Carry the material into the chamber
   - Air tight the door
   - Close the fresh air inlet valve

- Close the partial injector low valve
- Close opener cock
- Close gas sampling cock
- Set the pressure controller at required pressure
- Set up the recording alarm
- Switch on the vaporizer
- Switch on the caution lamp
- Start the vacuum pump
- After attaining the 50 per cent of required negative pressure, vacuum chamber to achieve complete sealing
- As the pump stops, close the exhaust valve immediately
- Put off the vacuum pump.
- Open the leak valve in order to return the inner pressure in the pump to normal and preventing the back current of pump oil
- Close leak valve

Can Application	Cylinder Application
• Set up the can in the holder	Connect the exclusive hose to application valve (upper) of partial injector
• Open partial injector low valve	Connect the hose to methyl bromide cylinder
• Open the opener cock	Open partial injector valve (low)
• Turn the opener handle of can opener and supply the fumigants into the chamber.	Open the top valve of cylinder and measure volume of liquid methyl bromide
• Close the opener cock	Open partial injector low valve, and confirm that finish application
• Take off the empty can from the holder	Take off the hose from the cylinder
• Switch off the vaporizer heater	Take off the hose from the

Switch off the ventilation fan injector Switch off the vaporization heater

Exhaust of Fumigants

- Open the opener cock again to exhaust the remaining gas in the partial injector
- Open slowly the fresh air inlet valve
- Switch on the agitation fan
- Switch on the ventilation fan
- Open the exhaust valve
- Operate the Vacuum pump, it should be on till the concentration reaches lower than 10 ppm inside the chamber
- Stop the vacuum pump
- Open the door. Before entering in the chamber again measure the concentration inside the chamber for safety.
- Switch off the ventilation fan
- Close the fresh inlet valve
- Close the can opener cock
- Close the partial injector low valve
- Switch off the agitation fan

Atmospheric Pressure Restored

Following the creation of the initial low pressure, atmospheric pressure may be restored in the chamber in several different ways, which may be summarized as follows:

(a) *Gradual restoration of atmospheric pressure:* The required dosage of fumigant is discharged and air is then slowly introduced until a pressure just below atmospheric is reached after 2 hours in a 3 hours exposure period.

(b) *Delayed restoration of atmospheric pressure:* Following discharge of the fumigant, the vacuum is sustained for about 45 minutes and the air is introduced rapidly into the chamber.

(c) *Immediate restoration of atmospheric pressure:* After the fumigant is discharged, atmospheric pressure is rapidly restored in the system by opening one or more valves leading into the chamber. This method, rather unscientifically described as the released vacuum or dissipated vacuum method, has been used extensively in North America for the fumigation of baled cotton.

(d) *Simultaneous introduction of air and fumigant:* In this technique, special metering equipment is provided whereby the fumigant is introduced simultaneously with air so that constant proportion of fumigant to air is maintained until the entire dosage has been introduced.

At the end of the treatment by any of these procedures, air washing is carried out as described above for the sustained vacuum technique. In a series of experiments using compressed bales of jute containing larvae of *Tenebroides mauritanicus*, it was found that the relative efficacy of these four methods from the view point of insect mortality was in the same order as listed. In effectiveness, the sustained vacuum technique occupied an intermediate position.

It must be emphasized that while the information given here may be useful as a guide for the investigation of vacuum fumigation techniques, each problem will have to be solved separately. The due consideration should be given to the commodity involved, the species of insects and their stages and the climatic factors in the country where the work is being done.

It is imperative to note that the some fumigants are going to be banned in near future for environmental safety. In this scenario, vacuum chamber may be used to make the seed lot insect free without using chemicals or fumigants. By altering the oxygen level

(0-0.5%) inside the vacuum chamber at the pressure 25-30 mm Hg, insect life can be killed under hypoxia. However, it requires several hours for complete mortality of insect life.

Classification of Fumigants

A large number of chemicals used as fumigants could be broadly grouped under three categories.

Liquid Fumigants: Chemicals having higher boiling point and exist in liquid form at normal room temperature, *e.g.,* EDCT mixture, ethylene dibromide, carbon disulphide etc.

Solid Fumigants: Chemicals available at tablets or powder form, react with atmospheric water vapour to release some toxic gas. e.g., Aluminium phosphide tablets, calcium cyanide powder etc.

Gaseous Fumigants: Chemicals having low boiling point and highly volatile at normal room temperature. Such chemicals are stored in metallic cylinders strong enough to withstand pressure exerted by the gas, *e.g.,* Methyl bromide etc.

a. Principal Fumigants

1. Ethylene dibromide or 1,2-dibromomethane ($CH_2Br.CH_2Br$)

The insecticidal properties of ethylene dibromide (EDB) were reported by Neifert *et al.* in 1925. It is a colourless, heavy organic liquid with a mildly sweet chloroform like odour with boiling point 131-132°C. The specific gravity is 6.49 (gas) and 2.17 (liquid) at 20°C and highly inflammable with a molecular weight of 187.88. It is one of the most toxic chemicals to insect damaging stored food grains. The use of EDB, filled in ampoules has been found suitable for fumigation of food grains stored in household structures, holding small quantities which can be made reasonably airtight. The fumigant is filled in glass ampoules and nozzles are sealed. Individual sealed ampoule is wrapped in cotton, blotting paper and then stitched in cloth bag. When ampoule with EDB liquid is broken and inserted into grain mass, the EDB changes into vapours. The ampoules are available in sizes of 3 ml, 6 ml, 10 ml, 15 ml and 30 ml. The dosage is 24-32 mg/lit with exposure period of 7 days in stack and 3 ml/q grain with exposure period of 3 days in small receptacles. It can be used in all food grains but not on milled products, oilseeds and moist grains.

Trade name: Orthofume, Bromotox, Fumogen, Bromofume, Escobrome-D, Dowfume W-85.

On account of low volatility, it is absorbed by the fumigated commodities, therefore, considerable aeration and long exposure are required for complete dissipation of vapours. It forms inorganic bromide in commodities containing fats and high moisture content. The absorbed portion in fat is slow to dissipate by ordinary aeration. On heating most of the fumigant is lost by volatilization. The small portion reacts with protein leading to formation of inorganic bromide. Shade and sun exposure, washing and cooking of fumigated grain have significant effect on the dissipation of residues. The MRL of inorganic bromide in India is as follows:

Food grain and milled food grain	- 25 ppm
Fruits	- 30 ppm
Dry fruits and spices	- 100 ppm

2. EDB$^+$

It is a mixture of EDB and carbon tetrachloride in the ratio of 1:8 w/w. It has a greater penetration power and hence can be used for bagged storage under gas proof covers also. It is equally effective in rural storage structures. EDB$^+$ is also available in ampoules in sizes of 11 ml, 22 ml and 30 ml for small quantities of food grains stored in rural storage structures. Smaller sized ampoules (1-5 ml) are available for disinfestation of even a few kg of grains and pulses. The dosage is 11 ml/q grain with exposure period of 7 days.

3. Durofume

It is a mixture of EDB and Methyl bromide in ratio of 1:1 w/w first prepared by CFTRI, Mysore. The mixture is highly effective for all stages of insects and even against resistant species of stored pests like *Tribolium* and *Trogoderma*. The dosage of durofume is 1-2 lbs/1000 ft^3 with exposure period of 48 hours. If temperature is above 26°C the exposure period may be reduced to 36 hours. If temperature is below 18°C the mixture may be altered to 1:3. For oilseeds, the mixture is in ratio of 1:2 and dosage is 1.5-2.0 lbs/ $100m^3$ with exposure period of 36-48 hours and for milled cereals 1:3 mixture is used @ 2-3 lbs/1000 ft^3 with exposure period of 48-72 hours.

Residues of mixture obtained on commodities were far below than that of EDB or MB used alone.

4. Hydrogen Cyanide or Hydrocyanic Acid (HCN)

It was introduced by American Cynamid Co. in 1783, first employed as fumigant by D.W. Coquillet in 1886 for cottony cushion scale, *Icerya purchasi*. It is a volatile, colourless liquid with a bitter almond odour and a Sp. gr, 0.699 at 20°C. The gas has Sp. gr. 0.9483 and is highly inflammable and explosive in mixtures above 5.6 per cent or above 4 lb/1000 cu. ft. It is one of the most toxic insect fumigants. HCN may be employed for fumigating many dry food stuffs, grains and seeds. In the market two forms of CaCN are available. Cyano gas contains 40 to 50 per cent CaCN and the gas is liberated slowly. The calcyanide dust contains 88 percent CaCN. For general fumigation, the dosage is 1 lb sodium cyanide (98 per cent), 1½ pints H_2SO_4 and 3 pints water per 1000 to 1500 cu.ft. It is available under the trade names of Calcid, Cyanogas, Cymog and Cyclon.

HCN does not react with the constituents of food stuffs of normal moisture content. Cereals absorb it during fumigation which is progressively reduced in the handling operation. In India, the MRL is as follows:

Food grain	-	37.50 ppm
Milled food grain	-	3.0 ppm

5. Carbon Bisulphide (CS_2)

CS_2 was one of the first fumigants employed on a large scale. Its use in France, in 1869, against the grape phylloxera is a landmark in the history of applied entomology. It was injected into the soil to control the insects infesting the roots of the grapevine. For many years afterwards, CS_2 was widely used as a soil or space fumigants. It is a colourless liquid with a disagreeable odour due to traces of hydrogen sulphide. Sp. gr. is 1.263 at 20°C and vapour pressure 360 mm Hg at 25°C. It is slightly soluble in water but soluble in organic solvents, highly inflammable and explosive in mixtures above 1 per cent in air. It is used for the control of stored product pest in godowns and pests of clothes, dress and draperies. The dosage is 10 lb of CS_2 per 1000 cu.ft. of space or 2 to 3 gallons per 1000 bushels of grain for 36 to 72 hr. The fumigant is highly phytotoxic.

The fat containing materials should not be fumigated as unreacted residues are likely to persist in such foods. This fumigant reacts with the peptides of amino acids of the food stuffs of the vegetable origin but does not lead to the formation of residues. Fat containing materials should not be fumigated as unreacted residues are likely to persist in such foods.

6. Carbon Tetra Chloride (CCl_4)

It was first employed as fumigant for the control of stored insect pests by A.P. Morse in 1908. It is a colourless liquid having b.p. 76.8°C and molecular weight 153.84 with Sp. gr. 1.595 at 20°C and vapour pressure 114.5 mm Hg at 20°C. It is almost insoluble in water but soluble in organic solvents. The gas has a Sp. gr. of 5.31, is not inflammable or explosive and possesses pungent chloroform like odour. It is comparatively less toxic to insects and slow acting. It has no effect on viability of seeds. The dosages are 2.5 lit./m^3 with exposure period of 14 days.

The fumigant is taken physically without chemical reaction in the grain and grain products. The fat containing foods may absorb large quantities and residues remain without chemical reaction. The chemical gets desorbed on aeration, movement and heating or cooking.

7. Ethylene dichloride or 1, 2-Dichloroethane, (CH_2ClCH_2Cl)

It was recommended by Cotton and Roark in 1927. It is a colourless liquid, chloroform like odour and gas being highly inflammable and soluble in organic solvents. Its boiling point is 83.5°C, mol. wt. 98.97, sp. gr. of gas is 3.42 and that of liquid 1.257 at 20°C and vapour pressure 78 mm Hg at 25°C. It is slow acting and death of insect takes place from one to three days after exposure and toxic to plants and human beings. It is available under the trade names of Brocide and Destruxol.

The absorption of the fumigant is higher at higher temperature. Handling, cleaning and milling process of fumigated grain helps in the dissipation of absorbed chemical.

8. EDCT or Ethylene Dichloride and Carbon Tetrachloride Mixture

It is a general mixture with 3 parts of ethylene dichloride and 1 part of CCl_4 by volume. This mixture was evolved from USA by United States Department of Agriculture. It is colourless liquid with chloroform like odour and both gas and liquid are non-inflammable. All kinds of cereals, pulses and millets have been safely fumigated at the dosage of 30-40 kg/100 m^3 with the exposure period of 48-72 hrs depending upon intensity of infestation, moisture percentage of the commodity, temperature and humidity.

The chemical present on the grains fumigated gets reduced by turning of the grain and the residues completely lost during cooking of the bread.

9. Phosphine (PH_3)

It was introduced as pesticide by A.G.Degesch in 1955. In India aluminum phosphide tablets of 3 and 1g, capable of liberting 1/3 in weight as phosphine are available under the trade names of Celphos, Quickphos Phostoxin, Phosfume etc. It produces carbide or garlic type odour. It is heavier than air and has low solubility in water. It is highly toxic to all stages of insect having good penetration power and also highly toxic to mammals. Aluminum phosphide is formulated as tablets, pellets or small sachets of powder in trilaminated pouch with additional materials such as ammonium carbonate (40%) and aluminum oxide (5%) to regulate release of fumigant and suppress flammability. It is used for effective control of common stored grain pests of cereals, millets, pulses, milled products, spices, oil seeds, oil cakes, bales and in other packages. The shelf life is two years. It is banned for public use, used by public agencies only.

There is no chemical reaction of PH_3 with the treated material at recommended dosage but the food stuffs with high moisture content will increase the total phosphorus content. The PH_3 after fumigation deabsorbed during cleaning and aeration. Shade and sun exposure, washing, cooking and preparation of *Dalia* and flour have significant effect on the reduction of residues. No ill effect on the organoleptic qualities of cooked product is observed. The MRL in India is as follows:

Food grain	-	0.05 ppm
Milled food grain	-	0.01 ppm

10. Eco_2 Fume

It is a product of M/S Cytee industries in USA. It is a non-inflammable, premixed cylinderized mixture of phosphine and CO_2, which provide highly effective fumigation in both sealed and unsealed storage facilities. Eco_2 Fume fumigant gas is a gaseous mixture of 2 per cent phosphine by weight in 98 per cent CO_2. It can be used in a wide variety of applications to attain insect free and dust free products. It doesn't affect real value of seeds. It can also be used to fumigate grains, dried fruits, cocoa, coffee beans, nuts, tobacco, cut flours and other items. It is well suited to fumigate structures and buildings.

11. Profume (Sulfuryl fluoride)

Sulfuryl fluoride (SF) possesses many positive fumigant qualities, including efficacy in controlling a broad spectrum of insect pests, a low boiling point, excellent penetration qualities, low reactivity potential, limited sorption, and rapid aeration. SF is inorganic, not combustible, has no flash point, and is very stable under most fumigation environments. It is a product of M/S Dow Agro-Sciences and is effective as methyl bromide on all postembryonic life stages of insects. It can be used for short or long (*i.e.* 4 -12 hrs) fumigation to meet the time constraints. Unlike methyl bromide, it does not deplete the ozone. Fumigated commodities less sorbs it. The gas is packed in cylinders and is non-inflammable non-corrosive, odourless and has a high vapour pressure and quickly penetrate the fumigated area.

12. Acrylonitrile

The compound is highly inflammable, been never used alone and always mixed with another suitable materials such as CCl_4 to reduce the risk of fire and explosion. The mixing ratio of acrylonitril and CCl_4 is 34:66. It is highly toxic to human being. It does not affect the viability of seeds and not leave any kind of residues in the fumigated commodity.

13. Grain-O-Cide

It is a mixture of CS_2 and CCl_4 in 4:1 ratio. The efficacy of mixture is directly proportional to its CS_2 content. The CCl_4 is mixed to reduce the fire hazard of CS_2. It has proved to satisfactory under Indian condition, and also acts as a useful soil fumigant. It is liposoluble and hence a nerve poison. It is used at a dosage of 30-40 kg/1000 q grains or 2.2 l/m^3 with exposure period of 7 days.

14. Chloropicrin (trichloronitromethane or nitrochloroform)

The chloropicrin (CCl_3NO_2) was first prepared in 1848 by Steinhouse. Its insecticidal use was reported in 1907 by Fabricwerke. It is a slightly yellowish liquid and 1.5 times heavier than water. The gas is powerful tear gas and 1.5 times heavier than air with molecular weight 164.39 and specific gravity of gas is 5.676 and that of liquid is 1.651 at 20°C. The boiling point is 112°C and non-inflammable with good penetrating effect. It is extremely toxic to insects, nematodes and certain fungi and it has wide application as a soil fumigant. The gas of this liquid has been found to reduce the germination power of seeds, therefore, should not be used for the fumigation of seeds. The dosage is 2 kg/1000 q grain exposure period 48 hours. It is available in the market under the trade names of Acquinite, Picfume.

No residues have been detected under normal conditions of fumigation. Formation of inorganic nitrites and nitrosamines has been detected. The fumigated grains should be thoroughly aerated as residues may affect milling and baking properties of the grain.

15. Methyl Bromide or Bromomethane (CH_3Br)

The insecticidal value of methyl bromide was first reported by Le Goupil (1932) in France. It is a colourless gas, b.p. 3.6°C, molecular weight 94.94, specific gravity: gas-3.27, liquid-1.732, non inflammable, solubility in water 1.34 g/100 ml at 25°C and soluble in most organic solvents and chloroform like odour. It is highly toxic to human beings even at low concentration. As the compound has no odour, chloropicrin is mixed @ 2 per cent as a warming gas. It has been extensively used in quarantine for plants, vegetables and some fruits, stored products, pests in soil, mills, warehouses, ships and railway wagons. The dosages are 24-32 g/m and 48 mg/lit.with exposure period of 48 hrs for stored grains and milled products, respectively. It has no effect on viability of dried seeds.

Repeated fumigation of seeds with methyl bromide should be avoided. Methyl bromide containing chloropicrin should not be used to fumigate plants, fruits or vegetables.

Under normal conditions, gaseous methyl bromide does not pose residue problem. It reacts with the proteins in plant tissues forming methylated derivatives and leaving residues of inorganic bromide. Absorption of the fumigant is relatively small and desorption through aeration is fast. It reacts mainly with the protein fraction of the food stuff. The use of methyl bromide at low temperature is more appropriate than other fumigants. It is largely absorbed in commodities having high fat content. Shade and sun exposure, washing and cooking of fumigated grain have significant effect on the dissipation of residues. In India, the MRL of inorganic bromide on food grains is 25 ppm.

Alternative Fumigants

Fumigation plays a major role in insect pest elimination in food grains during storage. Currently India uses about 470 tonnes of methyl bromide and 670 tonnes of aluminium phosphide for pest control in various stored products including food grains. Methyl bromide is a broad spectrum fumigant and the major advantage is its rapid action (exposure period 24-48 hours only) against the target pests. However, because of its ozone depleting potential, there has been a global move to phase out methyl bromide as per the Montreal Protocol 1992. In India, till July, 2000, the fumigant was used for treating the soil structures (flour mills) and stored commodities. Lately, the use of methyl bromide has been banned except for quarantine and shipment treatments.

Phosphine fumigant has several advantages (low cost, easy to apply, least residue and no ill effect on treated commodities). However, development of resistance in stored grain insect pests to phosphine in ware houses in the country has been reported to be a serious problem. Incidence of phosphine resistant strains of stored grain insect pests has been reported in several other countries as well. Thus the fumigants presently used are under pressure on environmental or biological grounds and there is a need to explore alternative fumigants to ensure continued protection of valuable food commodities. Realizing the significance of the problem, intensive research on carbonyl sulphide, ethane dinitrile, methyl phosphine and compounds from plant sources as alternatives has been carried out in different countries. The alternative fumigants, *viz.*, allyl acetate, ethyl formate, methyl phosphine, carbonyl sulphide, ethyl dinitrile, controlled atmospheres and plant products have been studied in India for controlling stored grain insect pests.

(a) *Allyl acetate*: Allyl acetate is a colourless liquid with acrid odour having the molecular weight 100.1 and boiling point 103.5°C. It is insoluble in water, highly inflammable and has low vapour pressure. It is toxic to human by ingestion, inhalation and contact, however, threre is no report indicating the compound carcinogenic. The rust red flour beetle, *T. castaneum* was the most tolerant to allyl acetate, whereas, the lesser grain borer was so susceptible that cent per cent mortality of all the developmental stages of the insect was achieved at the dose of 150g/m^3 and 50 g /m^3, respectively. In small scale fumigation, allyl acetate alone at the dose of 200 g/m^3 in 72 hours exposure proved effective in controlling stored grain insects in wheat.

(b) *Ethyl formate*: It is liquid fumigant having low boiling point (54.1°C) with a pleasant odour. It is naturally present in certain raw and processed foods at low levels (0.05-1.0 mg/Kg). it has rapid action against insect pests and known to break down into non-poisonous natural products, ethanol and formic acid. It is generally recognized as the safe fumigant. It has been used for individual package fumigation of dry fruits in Australia, South Africa and other countries. In Australia, Vapormate®, a formulation containing 16.7 per cent ethyl formate in liquid CO_2 has been developed for insect control in food processing facilities (dosage 420 g/m^3, 24 hour exposure). CO_2 serves the dual function of reduction in flammability risk of ethyl formate and synergistic effect. The effective doses ranged from 300-400 g/m3 with a 72 hour exposure period for treating different cereals, pulses, spices and dry fruits.

(c) *Carbonyl sulphide*: Carbonyl sulphide gas is an effective and safe alternative to phosphine. This has got a significant potential as a new fumigant for use as an alternative to phosphine. The fumigant properties of carbonyl sulphide, an industrial gas have been elaborately studied in Australia and it has been patented in 1993. It is moderately toxic and can achieve satisfactory insect control in a period of 6 hours to 7 days. This fumigant does not affect the quality of treated wheat and germination. It is likely to be registered as a grain fumigant in Australia under the trade name 'Cosmic'.

(d) *Ethane dinitrile*: Ethane dinitrile or cyanogens is a colourless gas with an almond like odour. It has got potential application for the control of wood borers, termites and stored grain insects. It is highly phytotoxic and hence it is useful for weed control and sterilization of grains. In laboratory tests at CFTRI, Mysore, it was noted that Khapra larvae and the developmental stages of rice weevil were highly tolerant and require 32.8 g/m^3 for 48 hours to achieve 100 per cent mortality. The fumigant is likely to be registered in Australia under the trade name 'Sterigas'.

(e) *Controlled atmosphere*: The controlled atmosphere system protects the grain quality and controls insects and mite pests. This referes to altering the normal atmosphere by introducing inert gases in sealed storage. The combination of inert gases with different fumigants needs to be tested with the objective to reduce the dosages of fumigants and to avoid various problems related with excessive use of pesticides. Complete mortality of all stages of resistant strains of *T. castaneum, O. surinamensis, E. cautella* and psocids was achieved in rice stacks following CO_2 treatment and there is no change in colour or odour of rice and wheat. It requires a long period of 10 days and above for disinfestation of food grains.

(f) *Plant products*: The compounds from plant sources may have the advantage over conventional fumigants in terms of low mammalian toxicity, rapid degradation and local availability. More than 75 native plant species belonging to different families have been tested in laboratory for fumigant toxicity.

The active components showing insect toxicity in the vapour phase can be grouped as monoterpenoids, cyanohydrins and cyanates, sulpur compounds, alkaloids etc. Essential oils or their components are generally less toxic to insects. When compared with conventional fumigants, they can be used only for small scale applications or for space treatments. In India, *Eucalyptus camaldulensis* Hook oil is suitable for stored grain insect pest control.

General Directions and Precautions for Fumigation

All fumigants are more or less toxic to human beings and domestic animals; hence, they must be used very carefully. The following consideration should be kept in mind while fumigating.

- Food stuffs are removed from the building to be fumigated.
- Edibles to be fumigated should be with proper safety to human lives, operation should be carried out by trained personnels with protective clothings.
- Enclosure must be airtight as possible. Avoid windy or cold weather.
- Use accurate dosages, proper exposure and ventilation for live plant fumigation.
- Optimum temperature for fumigation of live plants is 14-27°C.
- Optimum temperature for fumigation of stored grains/ receptacles is 21-30°C.
- Commodities containing fat should not be fumigated with EDB/MB.
- Study literature of fumigants before the operation.
- Keep first aid ready and wash hands after fumigation.
- Destroy empty containers.

- Put 'Poison Gas Danger' at the entry of building till proper ventilation.
- Call a physician immediately in accidental poisoning.
- Avoid smoking within 15 metres of fumigated area.

Chapter 13

Safety Measures Against Poisoning

All due precautions have to be exercised in handling pesticides because many of these leave toxic residues on treated food materials and plants which accumulate in higher concentrations in animals which consume these.

General Precautions

(a) Read the precautions given by the manufacturer of the pesticides carefully. Store the pesticides under lock and key; prevent access of children to these; empty containers must be disposed off safely.

(b) Blood and skin contaminations must be avoided.

(c) The godown for fumigation must be made airtight by closing the doors, windows, ventilators, crevices etc.

(d) In case of fumigation, the operator must use a suitable gas mask. It is a face-piece connected to a canister containing an absorbent material for the vapours of the fumigants by means of a flexible non-kinking breathing tube and provided with valves so that all air inhaled by the user passes through the canister and the exhausted air passes directly to the outside atmosphere through a non-return valve. The absorbent canister carries instructions indicating the number of hours for which it is effective in an atmosphere charged with the fumigant.

(e) After the required exposure period, aerate the godown thoroughly. First open the entrance door, keep it open for 6-7 hours, then only enter and open other doors or windows.

First Aid Precautions

- Before the doctor arrives or the patient is taken to the nearby hospital the following first-aid measures should be taken.

Inhaled Poisoning

- Remove the patient from source of contamination to fresh air and loose his clothings.
- Give artificial respiration
- Keep patient calm and comfortable
- Wrap the patient in a blanket to avoid chilling
- Do not give alcohol in any form
- In case of EDB/MB poisoning, the patient should be kept in sitting position and should not be allowed to sleep straight.

Swallowed Poisoning

- Remove poison by inducing vomiting
- Give common salt 15 g in a glass of warm water as an emetic and repeat until the vomit fluid is clear
- If patient is already vomiting do not give emetic with large amount of warm water.
- Vomiting should not be induced in case of the patient is in coma and convulsion
- In case of poisoning with CS_2, avoid vomiting instead give milk or egg white
- Administer 5 per cent glucose i/v in order to prevent collapse of the patient.

Skin Contamination

- Remove contaminated cloths and wash contaminated skin with stream of water and with soap.
- Rapid washing is most important for reducing injury.

Eye Contamination

- Hold the eye lids open
- Wash the eye with stream of running water immediately
- Do not use any eye drop until physician arrives

Prevention of Collapse

- Cover the patient with a light blanket
- Do no use a hot water bottle
- Raise foot of bed
- Give strong tea or coffee
- Administer stimulant like caffeine and epinephrine
- Administer 5 per cent dextrose i/v

Antidotes

a. General antidotes

(a) Remove poison by inducing vomiting

(b) Give 15 g of universal antidote in half glass of water. It consist of amixture of activted charcoal (two parts), magnesium oxide (one part) and tannic acid (one part).

(c) In case of sign of illness, *e.g.*, giddiness, nausea, headache, blurred vision, vomiting, depression, difficulty in respiration etc. a physician should be called or the patient be rushed to a nearby hospital. The doctor should also be told about the pesticide the patient had used.

b. Specific antidotes

(a) *Ethylene dibromide* : Liquid on skin may cause blisters. Inhalation causes pulmonary lesions and may lead to pulmonary oedema. Respiratory and circulatory depression may occur. Necrosis of the liver and kidney damage has been reported. The condition resembles chloroform poisoning and poisoning by other ethylene compounds, diseases of the central nervous system (CNS). Delayed pulmonary oedema, liver and kidney damage will have to be distinguished from other causes for these conditions. Oxygen therapy is desirable. If the poison has been swallowed, gastric lavage and sealing purgastives may be used. Respiratory stimulants like nikethamide or caffeine are useful. After acute nervous systems subside, adrenergic blocking agents like dibenzyl betachloroethyl amine hydrochloride has been recommended to prevent kidney and liver damage.

(b) *Methyl bromide*: Remove patient to fress air immediately and keep him warm. Administer oxygen therapy. Inject caffeine sodium benzoate 0.5 g. Keep the mouth rinsed always with 10 per cent solution of sodium thiosulphate.

(c) *EDCT mixture*: Sign and symptoms are produced mainly by central nervous system depression, resembling the effects of chloroform. There may be respiratory depression and delayed pulmonary oedema. Kidney and liver damage may occur. If swallowed, nausea followed by vomiting and abdominal pain, headache, dizziness, confusion, drowsiness and occasionally convulsion may occur. Visual disturbances may precede central nervous system depression. Oxygen therapy is desirable. If the poison has been swallowed, gastric lavage and saline purgatives may be used. Dimercaprol (British anti-levicide) may be tried against ethylene dichloride. Respiratory stimulants like nikethamide or caffeine are usefull. Steps should be taken to prevent liver damage. After acute nervous symptoms subside, adrenergic blocking agent like dibenzyl betachloroethyl amine hydrochloride has been recommended to prevent kidney and liver damage.

(d) *Carbon disulphide* (CS_2): Administer 5 per cent carbon dioxide in oxygen and inject 5-10 g of calcium gluconate.

(e) *Cyanides*: Hold before the nose of the patient amyl nitrite for 5 seconds at intervals of 15 seconds. Inject intravenously 10 cc of 3 per cent sodium nitrite solution at 2.5 cc per minute followed by 50 cc of 25 per cent sodium thiosulphate solution. Administer artificial respiration if breathing has stopped

(f) *Aluminium phosphide*: The poisoning symptoms are feeling of fatigue, buzzing in ears, nausea, pressure in the chest, intestinal pains, diarrhoea and vomiting. In medium to severe poisoning , dry cough, chocking attacks, intense thirst, reeling gait, severe pain in limbs, enlarge pupils and rapid set of stupor occurs. Oxygen therapy is desirable besides administration or a cardiac tonic and a drug to stimulate blood circulation. Under some circumstances blood transfusions or infusion of isotonic saline or glucose into the blood system are indicated. In the case of pulmonary edema a hypertonic solution of glucose be injected.

(g) *Strychnine*: Administer universal antidote followed by an emetic; emetic repeated until the vomit fluid is clear. Keep the patient warm.

(h) *Warfarin*: Inject large doses of vitamin K intravenously; also administer orally vitamin K and water soluble forms of Menadione. In case of large quantities taken for many days, fresh, whole blood transfusion is neceeary together with large doses of vitamin K_2

(i) *Zinc phosphide*: In case of poisoning within 24 hrs, administer a teaspoonful of mustard powder in a glass of warm water to induce vomiting; after vomiting has stopped, give 5 g potassium permanganate dissolved in a glass of warm water after 10 minutes.

Chapter 14

Insecticide Resistance Management

Insecticide resistance to both fumigants and contact insecticides is an increasing problem in controlling stored grain and grain product insects throughout the world (Table 14.1). Hence, persons engaged in food commodity storage and preservation world over are concerned about the resistance of stored product insects to contact insecticides and fumigants. It is, therefore, necessary to work out strategies to delay or minimize the probability of resistance evolution. Resistance to pesticides is the ability of a strain of insects to tolerate doses of toxicants which would prove lethal to a normal population of insects of the same species. Resistance results when occasional resistant individuals arise in a population and survive the pesticidal treatment. These survivors then reproduce and confer the resistance to their offspring in succeeding generations. Resistance is the ability of a population (of insects) to survive a dose (of insecticide) that would normally kill 100 per cent of a susceptible population. Insecticide resistance is a dynamic phenomenon dependent on biochemical, physiological, genetic and ecological factors.

The chances of development of insecticide resistance in stored grain insect pest are more because of their life style, *i.e.,* to remain confined in limited area. This results in more common inbreeding of the homozygous resistant strain resulting into homozygous resistant progeny. At the same time their management, once detected, can be easily achieved with the application of sufficiently acceptable sufficiently higher dose of the most appropriate insecticide to eliminate all the resistant survivors

The insect pest of stored grains can be conveniently divided into two groups, *i.e.*, coleopterous and lepidopterous, therefore, the subject on these two groups has been pooled accordingly.

Coleopterous Insect Pests

In a global survey, 11 species of coleopterous insect pests showed resistance to BHC, malathion, various fumigants, pyrethroids, cyclodines, DDT, carbaryl and juvenile hormone mimics. *Tribolium castaneum* showed resistance to BHC in 71 countries, and to malathion in 70 countries. *Oryzaephilus surinamensis, Rhizopertha dominica, Sytophilus oryzae; S. zeamaise* are resistant to g-BHC in more than 30 countries, and R. *dominica* was observed to show resistance to malathion in 25 countries. Out of 1 1 strains of *T. castaneum in* Malavi collected from cribs, 6 showed resistance to g-BHC and 3 to malathion.

Lepidopterous Insect Pests

Plodia interpunctella, ephestia cautella and E. *kuehniella* population showed resistance to malathion. Carboxyesterases and hydrolases contributed for non-specific resistance to organophosphates. The biology of malathion, and lindane resistant and normal (susceptible) strains of *Cadra cautella* were carried out at 30°C and 75 per cent RH. The incubation period, larval period in malathion and lindane resistant and normal strain were 2.7, 17.5, 2.9 and 22.6, 2.7 and 18.9 days, respectively. The pupal period in case of malathion and lindane resistant strains did not vary and it was 6.5 days for normal strain. The number of eggs per female were 158.9, 179.3 and 223.5 in Malathion and lindane resistant and normal strain, respectively. *E. cautella,* and *P. interpunctella* collected from groundnut stores in USA showed high and wide spread resistance to malathion. First report of *Bt* resistant field strain of *P. interpunctella* appeared from USA.

Table 14.1: Insecticidal resistance in insect pests of stored grains

Insect Pest	Insecticide	Country
Tribolium castaneum	DDT, malathion,	India
	Lindane	India
	Phosphine, dichlorvos	India
Sitophilus oryzae	Malathion, Lindane	India
	phosphine	India
Oryzaephilus surinamensis	Malathion	India
Trogoderma granarium	Phosphine	India
Plodia interpunctella	Malathion	USA

Types of Resistance

- Resistant insects may naturally detoxify or destroy the toxin faster than susceptible insects or quickly rid of their bodies to the toxic molecules (Metabolic resistance).
- The site where the toxin usually binds in the insect has been genetically modified to reduce the product's effect (Altered target-site resistance).
- Resistant insects may absorb the toxin slower than susceptible insect (Penetration resistance).
- Resistant insects may detect or recognize a danger and avoid the toxin (Behavioral resistance).

Resistance management programmes must be designed to retain a high level of susceptible genes in the pest population and thus maintain the effectiveness of insecticides. The most effective resistance management programme involves using insecticide alternatives and applying pest control measures or insecticides only when pest populations exceed economically damaging levels or thresholds. Using lower insecticide application rates, fewer applications, and less persistent pesticides can slow the development of resistance. Eliminating resistant pests with suitable non-insecticidal methods can increase the ratio of susceptible to resistant genes. If two insecticides are available for insect management, they can be used sequentially or as mixtures, in rotation. Integrated pest management programmes should be designed to minimize the development of resistance and movement of resistant individuals throughout the commodity marketing system.

Conclusion

a. General

- Wide spread cases of insecticide resistance in almost all the insect pests of storage to all the commonly used insecticides are available.
- There are evidences of development of cross resistance in many cases.
- Comparatively resistant problem is more pronounced in public bulk storage/flour mills as compared to farmers small storage because insecticides are more comminly used in bulk storage.
- Insecticide resistance is more common in coleopterous insect pests, more particularly in *T .castaneum, R. dominica S. oryzae* and *Oryzaephilus surinamensis* as compared to lepidopterous.
- Varieties may develop in the biology of resistant strains
- Abandoning of insecticidal pressure results in break down of resistance.
- The insect population shows susceptibility to the newly/ recently introduced insecticide.
- Inadequate airtight treatment facilities led to chronic under dosing and result into development of resistance and control failure.
- Most common detoxification mechanisms are MFOs, carboxyesterase, cytochrom P-450 etc.

b. Resistance management

- Use of non-chemical methods
- Introduction of susceptible strain. This approach is not feasible in case of storage
- Use of acceptable high dose of the most appropriate chemical to eliminate the resistant strains
- Use of combinations or alternate insecticides
- Regular monitoring to have an account of resistance development
- Maintain hygienic condition, do not add new stock to the treated residues.
- Coinciding the treatment with the susceptible pest stage.

Chapter 15

Integrated Pest Management in Stored Products

The produced food grains have to be stored either for short or long duration for utilization in the coming times for food, feed and seed. In storage, these undergo 9.33 per cent losses due to various agencies such as rodents, insects and mites, microorganism and moisture in India, whereas, in other developing countries, the losses are as high as 30 to 50 per cent. Therefore, with an urge to produce more, there is still urgent need to protect the produce in the storage to meet the challenge. As a result of improved storage technology, the losses in Government Warehouses could be reduced to minimum but at farmers' level, where 60-70 per cent of the produce is retained, the losses are still too high.

The management of insects in stored products is in transition from a dependence on regular applications of chemical insecticides to the use of integrated pest management (IPM). IPM involves understanding interactions between stored product environment and insects associated with stored products, and replacing all or most of the chemical applications with cost effective non chemical alternatives. The increased emphasis in adopting IPM strategies for stored product insects is primarily due to consumers' demand for food free from pesticide residues. The transition from a heavy reliance on chemicals to little or no chemical input and increased use of non chemical approaches requires a thorough understanding of stored product insects, their identification, monitoring, biology, ecology and response to chemical and non chemical management options. The current scenario of stored grain protection in India depends totally and judicious use of IPM strategies. For safe storage of food grains, both the preventive and curative measures are required:

1. Preventive Measures

'Prevention is better than cure', a common proverb also implies here for management of insect pests of food grain. Preventives measures recommended are as follows:

(i) *Hygiene and sanitation:* The various factors responsible for food grain losses in storage considerably can be reduced to minimum by maintaining hygiene and sanitation in godown. If the stocks are kept in neat stores in scientific manner, a good percentage of food grain loss can be reduced. Before storing the grains in an ideal storage/godown, it is important to ensure that the food grains are dry, clean, cool, wholesome and free from any obnoxious odour. The storage structures should also be neat and clean and reasonably air tight so that pest control measures may be possible. The following hygiene and sanitation measures may be adopted.

(a) Threshing floors or yards should be clean, free from any insect infestation, away from the vicinity of village or granaries to avoid horizontal and cross infestation, as most of these insects can move from these sources to infest grains on threshing floors or yards

(b) Harvesting and threshing machines and different means of transport like truck, trolleys or bullock carts etc. should be cleaned before their use to avoid the risk of latent infestation of storage insects.

(c) Cleaning of storage structures, receptacles and godowns before storing newly harvested and processed crop.

- Remove all dirt, rubbish, sweepings and webbings etc. from the storage structures and stores and dumped or destroyed.

- All the cracks, crevices, holes, walls and ceiling should be closed with cement permanently.
- All the rat burrows should be closed with a mixture of broken glass pieces and mud and then plaster with mud or cement.
- White wash the stores before storage of food grains and try to make it rodent and moisture proof.

(d) The used bags are the main source of infestation and these should be treated properly with some insecticide or fumigant to check the latent and cross infestation.

(ii) *Proper stacking :* Proper stacking of the bags also helps in grain protection. Stack properly the stitched bags on the dunnage (wood/plastic) 0.5 metre away from the walls and floor in the lines keeping 0.5 metre space between two lines and 0.2 metre open space towards the roof. Do not stack more than 10 bags vertically.

(iii) *Disinfestation of storage containers, structures and stores :* Disinfest the storage containers and structures before storing of the grains. This will not only kill the latent and hidden infestation of the insects but also check their cross infestation from other bags and storages.

(a) Clean and disinfest empty used gunny bags by dipping in the boiled water and drying under sun, when re-used. If one can afford, treat old gunny bags by dipping in cypermethrin 0.01% or fenvalerate 0.01% or malathion 0.1% for 10 minutes and dry them completely under sun.

(b) Treat all the inner surfaces and walls of the storage structures and godowns either by malathion (50 EC) or fenitrothion (50 EC) using 5 ml of any of the insecticides in one litre of water or deltamethrin (2.5% WP) using 1.25 g in one litre of water by spraying this aqueous solution at the rate of 20 ml/m^2 area or fumigate with aluminum phosphide @ 21-30 g or 10.0 litre EDCT/1000 feet^3. Dry the treated structures completely before storing the grains.

(iv) *Legal method :* Entry of an insect which is not found in the particular area, can be prevented by the imposition of Destructive Insect Pest Act, 1914.

2. Curative Measures

- Ecological
- Mechanical
- Physical
- Cultural
- Botanicals
- Chemical
- Biological
- Engineering
- Other methods

(i) *Ecological measures*

The safety of stored grain from insect pests largely depends on the proper management of three factors, *viz*.,

(a) Storage temperature

(b) Grain moisture content

(c) Availability of oxygen

All these factors are required for rapid development and multiplication of insects. Hence, they have to be properly manipulated through design and construction of storage structures or godowns and storage practices so as to create ecological conditions unfavorable for attack by the insects and mites.

(a) *Storage temperature:* Temperature ranging from 25-35°C accelerates the development of the insects and above 40°C and below 15°C retards reproduction and development, while temperature above 45°C and below 10°C kills the insects. Heating of the grains to 50°C may be lethal to insects but it is not advisable because the grains may loose their viability.

(b) *Grain moisture content:* Grains stored around 10 per cent moisture content escape from the attack of insects. However, khapra beetle can develop even below this moisture content. Over the moisture range of about 12-16 per cent, the rate of seed deterioration increases as the moisture content increases. The losses could be rapid due to mould growth and/or due to heating.

(c) *Availability of oxygen:* The reduction in oxygen level in the grain containers is mainly due to the respiration of grains and also by the insects. The reduction in oxygen level to a critical limit stops further multiplication of a particular species. The khapra beetle, which can thrive even at much lower moisture content, is very susceptible to the reduction in oxygen content. Reduction of O_2 availability can be achieved by addition of CO_2 or N_2.

(ii) *Mechanical control measures*

Mechanical control methods are quite practicable which have been discussed below:

a. Screening

- Screening of the grains is important. Broken and cracked grains promote attack of storage insects. Hence, screening or sieving out such grains reduces the insect infestation and gives a better appearance to it.

- Regular screening of grain should be done away from the stores to avoid any reinfestation.

- Screenings should be destroyed immediately.
- Bags used for carrying the screenings should not be used again unless disinfested.

b. Grain protection devices

- TNAU Insect Probe Trap
- TNAU Pitfall Trap
- TNAU Two-in –one Model Trap
- Indicator Device
- TNAU Automatic Insect Removal Bin
- UV-Light Trap for Grain Storage Godown

All these devices can be used for both monitoring and mass trapping of stored grain insects. It is important to note that presence of even a single live insect in the food grains can not be tolerated as it may build up and cause enormous loss in storage due to its high reproductive potential.

TNAU Insect Probe Trap

Use of trap is relatively a new method of detecting / trapping insects in stored grains. The basic components of a TNAU probe trap consist of three important parts: main tube, insect trapping tube and a detachable cone at the bottom. Equispaced perforations of 2 mm diameter are made in the main tube.

Concept : Insects love "AIR" and move towards air. This behaviour of the insect is exploited in this technology.

Method of working : The insect trap has to be kept in the grain like rice, wheat etc., vertically with the white plastic cone downside. The top red cap must be kept to the level of the grain. Insects will move towards air in the main tube and enter through the hole. Once the insect enters the hole it falls down into the detachable white cone at the bottom. Then there is no way to escape and the insects are trapped for ever. The white detachable cone can be unscrewed once in a week and the insects can be destroyed.

Salient Features : No chemicals; no side effects and no maintenance cost.

Efficiency : TNAU Insect traps are excellent insect detection devices in food grains and more effective in the detection of stored grain insects namely *Rhyzopertha dominica* (F), *Sitophilus oryzae* L. and *Tribolium castaneum* (Herbst) in stored food grains both in terms of detection as well as number of insects caught than the standard normal sampling method (by spear sampling). The detection ratio (trap: normal sample) is higher in the trap than that of the normal sampling method by factors ranging from 2: 1 to 31: 1. The insects catch is also higher in the probe trap than the normal sampling method by factors ranging from 20: 1 to 121: 1.

These are also considered as good mass trapping devices when used at 2 - 3 numbers / 25 kg bin (28 cm diameter and 39 cm length). They should be placed at top 6 inches of the grain, where the insect activity is seen during early period of storage. They can remove more than 80 per cent of the insects within 10-20 days.

TNAU Pit Fall Trap

Pitfall traps are used for capturing insects active on grain surface and in other layers of grain. (Monitoring and mass trapping tool).

Standard Model

- Standard model of pitfall trap has 2 parts, perforated lid (2 mm (or) 3 mm) and cone shaped bottom portion.
- Application of special coating with sticky material on the inner side of cone to hold trapped insects is necessary
- This procedure is tedious.

TNAU Model

- TNAU model has perforated lid, cone shaped bottom which tapers into a funnel shaped trapping tube.

- Hence sticky coating is dispensed with
- Commercial model is in plastic, simple and economical (cost per trap is Rs. 25/- only).
- Easy to handle.

TNAU Two-in-one Model Trap

The probe trap containing the components namely the perforated tube, pitfall mechanism, a collection tube and the cone shaped pitfall trap with a perforated lid and the bottom tapering cone were combined as a single unit. Combination of probe and pitfall increase the trapping efficiency of insects. This trap is suitable for pulse beetles as they are seen only on grain surface wandering here and there. It does not require tedious procedure like coating the inner surface of pitfall cone with sticky materials before trapping to hold pulse beetles. Beetles are captured alive with this trap, which may facilitate release of pheromone and thereby attract more number of insects.

Indicator Device

It consists of a cone shaped perforated cup (3mm perforation) with a lid at the top. The cup is fixed at the bottom with a container and circular dish which are to be smeared with sticky material like vaseline.

Farmers, before storing their pulses, should take 200 g of pulses to be stored and put them in the cup. When the field carry over beetles start emerging, due to their wandering behaviour, they enter the perforations and get slipped or and fall into the trapping portions. As they stick on to the sticky materials, farmers can easily locate the beetles and take out them. stored pulses for sun drying. The device with 2 mm perforations ca: be used for cereals, too.

This will help in eliminating the initial population, which acts as the major source for further build up. Thus, timely detection will help the farmers to preserve their valuable pulses during storage. The device is being popularised.

TNAU Automatic Insect Removal Bin

TNAU insect removal bin can remove the insect automatically. The structure has 4 major parts, namely outer container, inner perforated container, collection vessel and the lid. The model exploits wandering behaviour of stored product insects as well as the movement of these insects towards well aerated regions. The grains are held in the specially designed inner perforated container. The space between inner and outer container provides good aeration for the insects. Insects, while wandering, enter the perforation to reach the aerated part and while doing so, get slipped into the collection vessel through a pitfall mechanism provided in the collection vessel. In order to collect the insects quickly, as and when they emerge from the grains, perforated rods (2 mm) are fixed in the inner container.

The container will be useful for storing rice, wheat, broken pulses, coriander etc. The insects such as rice weevil, lesser grain borer, red flour beetle, saw toothed beetle, which are commonly found attacking stored grains can be removed automatically by storing the grains in this.

Efficiency : Grains (paddy and sorghum) stored in Automatic insect removal bin (100 kg and 500 kg) recorded only 1-4 per cent damage by insects compared to 33 to 65 per cent damage in the ordinary bin after 10 months of storage. The population of insects *(R. dominica* and S. *oryzae)* ranged from 0-2/ kg in the grains stored in 100 kg Automatic insect removal bin compared to 5-191/kg in ordinary bin after 10 months of storage.

Significant : No granaries can be filled with grains without insects as the harvested produce contain, egg (or) larvae (or) pupae because of field carry over infestation which cannot be avoided in developing countries like India. So the aim of Storage Entomologist should target this field carry over population (which is generally < 1 %) from which adults emerge at 15 to 20 days after harvest in granaries. If this field carry over population is not cared, then the insect will multiply and contaminate the food grains and complete damage can even be expected within 4 months of storage. The automatic insect removal bin can remove substantial number of emerging adults and sometimes free living larvae also. Significant success has been achieved in popularisation of Automatic Insect Removal bin Technology.

c. Use of entoleter

Entoleter is a mechanical device in which infested grains or their products are revolved at a speed of 3000-4000 rpm, which destroys the insects by centrifugations of their haemolymph. This device is used by roller flour mills and some rice mills but its operation is not economical on rural storage system.

(iii) Physical control measures

(a) *Heat treatment:* Most of the stored grain insects die at 55-60 ^{0}C within a period of 10-20 minutes. Therefore, flour-mills and other types of food processing plants have arrangements of super heating by infra-red heaters for control of the insects.

(b) *Controlled atmosphere:* It is one of the safer storage practices. Controlled atmosphere (CA) disinfestations technology implies the alteration of natural storage gases, *i.e.* carbon dioxide, oxygen and nitrogen to render the atmosphere in the stores inimical to pests. CA also refers to the process of changing the composition of existing atmosphere artificially by introducing CO_2 or N_2. Controlled atmosphere may be achieved in several ways:

- Adding gaseous or solid CO_2 (High CO_2 atmosphere).
- Adding a gas of low O_2 content (*e.g.*, pure N_2 or output from hydrocarbon burner) Low O_2 atmosphere
- Allowing metabolic processes within a storage to remove O_2, usually with associated release of CO_2 (Hermetic storage atmosphere).

This artificial atmosphere prevents multiplication of insects, mould growth and quality deterioration of the food grains.

The storage atmosphere can be modified either by lowering oxygen concentration as 2-4 per cent by purging the nitrogen or increasing CO_2 concentration as 9-10 per cent by addition of dry ice (solidified CO_2), which are lethal to storage insects.

(c) *Mixing of inert dusts:* Some inert dusts are abrasive and scar the cuticle causing desiccation and starvation, but actual efficacy depends on the following factors:

- *Particle size:* Finer the particles, higher the insecticidal efficacy, subject to a maximum of 10 μ because coarser particles do not stick to the insect or the grains. Certain dust are, however, ineffective even if ground finely.
- *Particle hardness:* Generally, efficacy of dust increases with the intrinsic hardness of their particles.
- *Particle shape and form:* Hard crystalline and angular substances are effective.
- *Humidity:* Efficacy of dusts slows down at 70 per cent relative humidity or more.

The farmers in villages also mix send, clay ash etc. with the grains which make the entry of insects in grains a difficult task and causes physical injuries; blocks mouth parts, lodges in the joints of the insects. Ash or any other fine dust also blocks the spiracles of the insects and results in death due to suffocation. The principal mode of action of inert dusts is that they cause insects to desiccate. Insects die when they lose about 30 per cent of their total body weight. Some of the dusts made from silica aerosols, activated carbons operate by abrasive action on the waxy layer of insect cuticle, resulting in desiccation and death of insects. Food grains stored with inert dusts are to be cleaned and washed before consumption.

(d) *Solar energy:* Utilization of solar energy has long been practiced as one of the effective method for minimizing pest infestation in stored grains (Dean, 1911 and Thompson, 1915). The use of heat energy was first practiced in Australia to disinfest wheat by exposing the grains for at least 3 times at a temperature range of 60–66°C (Winterbottom, 1922). Mookherjee *et al.* (1968) observed a solar heat temperature range of 45-55°C to eliminate the infestation of *C. maculatus* in moong, pea and gram. Dortbudak (1975) reported that exposing containers of lentil and black gram infested with *C. chinensis* to sunshine for 4 hours (from 9.00 to 13.00) inhibited the rate of pest multiplication and reduced grain loss. Saxena *et al.* (1992) observed the effect of exposing 1,2 and 3 days old pupae of *T. granarium T. castaneum* and *C. chinensis* to temperatures of 30°C, 40°C and 45°C for 24, 48 and 72 hours. Observation on adult emergence and their progeny revealed that, although all the temperatures were effective in causing sterility, exposure at 45°C was particularly effective in suppressing adult emergence of *T. castaneum* and *C. chinensis*. Sharma *et al.* (1997) exposed the pupae of *R. dominica* to temperatures of 40°C and 45°C for 24-72 hours and reported that the adult emergence was totally suppressed at 45°C for 72 hours exposure. Heating at 45°C completely prevented hatching in one day old eggs of *R. dominica* but longer heating was needed for older eggs. Singh (2000) evaluated the effect of solar energy *vis-à-vis* colour of the storage bag by studying the biology of *C. chinensis* on green gram. It was reported that both black and blue colours were effective in harnessing the solar energy for *C. chinensis* control; 16 hours exposure was enough to cause egg mortality while 24 h exposure for grub mortality. Yadav *et al.* (2002) conducted experiment to harness solar energy and to evaluate the effect of different colour of polyethylene storage bags (black, blue, gram and yellow) for the management of *R. dominica* in stored wheat. They reported that the mortality of test insect increased while seed germination decreased with the increase in exposure duration. Maximum infestation was eliminated by exposing the infected grains to solar radiation for 24 hours in black polythene bag. Chander (2003) reported that exposure period of 24 hrs to solar energy in black and blue coloured bags was most effective to reduce infestation and adult emergence and to cause adult mortality of *R. dominica*.

(iv) Cultural control measures

(a) *Splitting of pulses:* Some insects like pulse beetles prefer whole pulses and not split ones. Thus split pulses escape from their attack and as such most suitable for storage than whole grains.

(b) *Parboiling of paddy:* Parboiling of the paddy is a common practice in rural areas in Eastern and South India and people prefer to eat parboiled rice than raw rice due to its taste. But parboiled rice is generally least susceptible to insects attack as compared to raw rice because of its hard glutinized outer layer.

(v) Botanical Control Measures

Traditionally, many of the plants are used against stored grain pests. Although, promising results have often been achieved in laboratory tests with plant material (botanicals), the efficacy under practical storage conditions varied as lot. Most methods have a limited effect but it provides satisfactory protection of the stored products when they are applied properly. There are many plant species, which have toxic properties against insect pests of storage. The most commonly used plants are *Neem, Karanj, Undi,* castor, lemon grass, mustard, coconut, sesame, groundnut, *Dharak, Aak, Datura* etc. The plant part extract, oil and plant part powder are used to admixture with stored commodities to protect them from insects.

Jotwani and Sircar (1965) found that the powdered *Neem* seed kernel when mixed with wheat seed at 1 to 2 parts per 100 parts of seed, protects it against, *S. oryzae, R. dominica* and *T. granarium* for at least about 269, 321 and 379 days, respectively. The germination of treated seed was not impaired. The efficacy of some indigenous plant products against infestation of *S. oryzae, T. granuraim* and *C. chinensis* showedthat powder of *Acorus calamus* rhizome was the most promising against the three pests exhibiting a fairly high protection of grains even at the lowest concentration of 1 per cent. Jacob and Sheila (1990) reported that *Neem* oil at the rate of 1.0 ml/100 g seeds caused cent per cent mortality of adult beetles of *C. chinensis* and inhibited oviposition and hence prevented development of subsequent generations. *Neem* oil at 0.5 ml/100 g seeds did not inhibit egg laying but prevented the development of progenies in the next generation. The vegetable oils *viz.,* sunflower, castor, mustard, safflower, groundnut, palm, sesamum, *Neem*, *Karanj* and corn (maize) were tested for ovipositional preference and egg hatching of *C. chinensis* on pigeonpea all the treatments had an adverse effect on ovipositional preference compared with the control. *Neem* oil at a concentration of 1.0 per cent was superior to all the other treatments. The egg hatch inhibition of *Corcyra cephalonica* increased with the increase in concentration of plant extracts. The LC_{50} values for egg mortality were 0.7079 ± 0.0581, 0.08954 ± 0.0566, 1.1041 ± 0.0597 and 1.4125 ± 0.0547 in lemon grass oil, *Neem* extract, *Karanj* extract and *Undi* extract, respectively. Against *C. chinensis* in cowpea, the castor oil @ 1.0 ml/100 g seeds was most effective in inhibiting the oviposition (26.6 eggs/female) as against 79.4 eggs/female in untreated seeds. At 1.0ml/100g seeds, castor oil caused maximum reduction in egg viability (61.7%) followed by mustard oil (56.7%). The longevity of male and female adults decreased with the increase in doses of oil. The treatment of castor oil @ 1.0 ml/100 g seeds caused maximum reduction in adult emergence in F_1 generation (85.0%) followed by mustard oil (83.7%), groundnut oil (73.3%). No adverse effect of tested oils was observed on the germination of cowpea seeds upto 150 days of treatment.

(vi) Chemical control measures

Among the present methods of control of stored grain pests chemical control is the most popular and the most effective one. Surface spray of bags as and when the insect pests are visible are done with insecticides such as malathion, primiphos-methyl, DDVP, fenitrothion etc. Malathion 50 EC is sprayed over the bags in the ratio of 1:100 @ 3 l/100 m^2. Recently the deltamethrin 2.5 WP @ 3 mg a.i. /m^2 has been found very effective. Allay charging or spraying of allays between two grain stocks with chemicals like dichlorovos (DDVP) @ 0.05 per cent has been found very effective in preventing the cross infestation.

Dust formulations of the insecticides are sold rarely for use and contain 0.1 to 5.0 per cent active ingredient. The formulations contain additives, which increased the adhesion to the stored produce. Dust formulations are suitable for mixing with grain and for applying in layers in the stored produce (sandwich method) as well as for surface treatment of individual bag, bag stacks and stores.

Fumigation is a process of creating lethal concentration of a toxic gas for the time necessary to destroy insect iinfestation in a closed space. If there is any sign of infestation in the grain, whether stored in receptacles or in godowns, it is best to fumigate it with one of the poisonous gases available for the purpose and then to sieve and dry it before returning to the receptacles. To ensure a complete mortality of the infesting pests, it is advisable to repeat the fumigation after one month. The details of fumigants, their use and precautions have already been discussed in the chapter 'Fumigants, properties and residues in stored agroproducts'.

(vii) Biological control measures

Biological control of insect pests in storage godowns is not very encouraging. However, some of the hymenopterous parasites such as *Anisopteromalus calandrae* on *Sitophilus* spp., *Lariophagus distinguendus* on *Rhizopertha* and *Sitophilus* have been recorded parasitising various lepidopterous pests of grains. A detailed account of predators and parasites has been presented in (Table 15.1). *Bacillus thuringiensis* has also been tried on various lepidopterous pests of grain stores. It is the first microorganism to be approved for use on stored grain pests. B.t. is compatible with most protectants, fumigants and seed fungicides. However, its wide use is yet to take up even in Western countries.

Table 15.1 : List of predators and parasites attacking stored pests

S. No.	Predator/Parasites	Attacking stage	Insect pest
	Parasites		
1.	*Aximopsis javensis* Girault	Larva	*Araecerus fasiciculatus*
2.	*Anisopteromalus calandrae* (Howard)	Larva	*A. fasiciculatus*, *L. serricorne*, *R. dominica*, *Callosobruchus* spp. *S. oryzae*
3.	*Antrocephalus aethiopicus* Masi	Larva	*C. cephalonica*
4.	*Antrocephalus mahensis* Masi	Larva	*C. cephalonica*
5.	*Aximopsis tephrosiae* Girault	Larva	*Araecerus fasiciculatus*
6.	*Cephalonomia gallicola* Ashmeed	Larva	*S. paniceum, L. serricorne*
7.	*Cephalonomia tarsalis* Ashmeed	Larva	*Oryzaephilus surinamensis*
8.	*Choetospila elegans* Westwood	Larva	*L. serricorne*, *R. dominica*, *Callosobruchus* spp., *S. granarius*, *S. oryzae* and *Trogoderma granarium*
9.	*Dimachus discolor* Walk	Larva	*S. paniceum*
10.	*Entedon* spp.	Larva	*S. paniceum*
11.	*Eurytoma tylodermatis* Ashmeed	Larva	*Araecerus fasiciculatus*
12.	*Habrobracon hebetor*	Larva	*Plodia interpunctella*
13.	*Habrobracon brevicornis* (Wesmael)	Larva	*C. cephalonica*, *Ephestia cautella*
14.	*Lariophagus distinguendus* (Forester)	Larva	*S. paniceum*, *L. serricorne*, *R. dominica*, *S. granarius* and *S. oryzae*
15.	*Pteromalus cerealellae* (Ashmeed)	Larva	*Sitotroga cerealella*

(Cont.)

16.	*Pteromalus pyrophilus* Kollar	Larva	*Sitotroga cerealella*
17.	*Ptinobius texanus* Crawford	Larva	*Araecerus fasiciculatus*
18.	*Pyxinia crystalligera*	Larva and Adult	*Dermestes* sp.
19.	*Trichogramma pretiosum* (Riley)	Egg	*Ephestia cautella, Plodia interpunctella*
20.	*Trichogramma cacoeciae* Marschal	Egg	*Sitotroga cerealella Corcyra cephalonica*
21.	*Venturia canescens*	Larva	*Ephestia cautella, Plodia interpunctella*
22.	*Zatrophus incertus* (Ashmeed) L.	Larva	*S. oryzae*
	Predatory mites		
23.	*Acarophenax* sp.	Egg	*Dermestes* sp.
24.	*Cheyletus* sp.	Egg and young larva	*L. serricorne, R. dominica*
25.	*Pyemotes ventricosus*	Egg	*Callosobruchus* spp., *T. granarium, Tribolium castaneum*
26.	*Pyemotes tritici*	Egg and young larva	*L. serricorne,*
27.	*Tenebroides mauritanicus*	All stages	*R. dominica*

(viii) Engineering control measures

Hermetic or air tight storage : Food grains kept in airtight sealed structures remain insect free, reason being that during storage the grain respire and liberate carbon dioxide. Excessive CO_2 accumulation and depletion of oxygen do not allow the insects to survive. Under ground storage of paddy in Eastern India is hermitic storage. Seeds cannot be stored by this storage method due to loss of viability.

(ix) *Other Methods*

(a) Effect of Different Storage Receptacles

Awaknavar *et al.* (1989) used storage structures, *viz.,* metal bin, coaltar drum, plywood bin, earthen pot and gunny bag at farm level for storage of pulses against *C. chinensis* and stated that the performance of metal bin and coaltar drum was better compared to other structures with reference to insect infestation with a loss in weight of 3.2, 4.2, 4.8, 6.3 and 5.0 per cent in metal bin, coaltar drum, gunny bag, plywood bin and earthen pot, respectively. Senguttuvan *et al.* (1993) studied the effect of different storage containers on the incidence of *Corcyra cephelonica* in stored groundnut both under artificial and natural conditions. They reported significantly low dry mass loss and damaged kernels in gunny bag under artificial condition followed by polyethylene bag and urea bag. Higher damage was noticed on kernels stored in mud pots. Under natural condition, low dry mass loss and damaged kernels were observed in gunny bag and urea bag. Chander (2003) noted significantly higher dry mass loss (8.53%) and damaged grains (32.43%) due to *R. dominica* when stored in mud pot in artificial condition while under natural condition, low dry mass loss and damaged grains were recorded in urea bag followed by gunny bag, polythene bag, mud pot and cloth bag. Meena and Bhargava (2003a) studied the effect of different storage containers on the incidence of *C. cephalonica* in stored groundnut under both artificial and natural conditions and found significant differences in damage on kernels stored in different containers under artificial and natural conditions. Significantly higher dry mass loss (5.29%) and damaged kernels (29.24%) was noticed in mud pot stored kernels. The number of adults varied from 2.33 in gunny bag to 4.00 in cloth bag. Under natural condition, low dry mass loss and damaged kernels were noticed in gunny bag followed by urea bag, cloth bag, mud pot and polyethylene bag.

Table 15.2: Radiation induced sterilization of insects

Insects	Stage	Radiation	Sterilizing dose (rad)	References
Acanthoscelides obsoletus	A	g	20000	Pesson, 1963
Attagenus grandis	A	g	15000	Davich and Lindquist, 1962
Callosobruchus chinensis	A	g	42000	Quraishi and Metin, 1963
Ephestia cautella	P	g	>16000	Pendlebury *et al.* 1962
Ephestia kuhniella	$P^?$	g	60000	Pelerents, 1963
Lasioderma serricorne	A	g	16000	Pendlebury *et al.* 1962
Plodia interpunctella	-	g	>16000-50000	Pendlebury *et al.* 1962
Rhyzopertha dominica	A	g	12000	Pendlebury *et al.* 1962
Sitophilus granarius	A	X	7500-10000	Hoover *et al.*, 1963
Tribolium castaneum	A	X	6000	Erdman, 1962
		g	16000	Crook, 1962
Trogoderma granarium	A, $P^?$	g	15000	Harvey, 1963 and Kanso, 1962

A= Adult
P= Pupa

(b) Irradiation

With the discovery of X-ray by Roentgen in 1895, there have come a long way to understand the mechanism of radiation in living organism. Various types of ionizing radiations used are: gamma rays, X-rays, ultra violet rays, alpha rays and beta rays. The ã and X-rays have great penetrating power and are most commonly used for insects. Irradiation of food grains with gamma rays is effective against stored grain insect pests as it causes mortality as well as sterility in insects depending on the dose and time of exposure (Table 15.2). The infrared rays and microwaves are also used in raising the temperature of store that disinfect the grains. The irradiation is measured either in *rad* (1 rad=100 erg/g or 10^{-2} Jule/kg of irradiated object) or in *Gray* {(Gy) (1Gy=100 rad)}

In insect control the lethal and sterilizing effect are produced by g-ray mainly from Co^{60}. It is possible to employ g-rays in doses lethal to the population or in doses, which sterilize the irradiated insect. In general, with increasing dosage of radiation the mortality of exposed insect increases and with further increase in dose the mortality takes place in a shorter time. A dose between 200-500 krads brings about complete mortality of stored grain insects. The egg and larval stages of insect are more susceptible to irradiation as compared to pupal and stage.

Effect of radiation on insects : When an insect is exposed to ionizing radiation, it brings about many harmful effects, *viz.,* effect at cellular level (increased membrane permeability, lysosomes rupture, alteration in DNA & RNA and death of intercellular structures like mitochondria), effect on embryonic development (metaphase suffered and death of embryo), morphological abnormalities (deformities of head, body segments, wings and other appendages and difficulties in moulting), effect on reproduction (brings sterility, reduction in fecundity and fertility and damage to oocytes or nurse cells), histopathological effects (damage to somatic tissues of mid gut, loss of peritrophic membrane and affect production of enzymes and hormones) and biochemical effects (inhibition of DNA synthesis in developing eggs, the alkylating azirinyl compounds react with DNA, RNA and protein and inhibition of malic acid, proteins and free amino acid in eggs).

c. Sound Waves

Very intense sound may carry sufficient energy to kill the insect. The death occurs due to heating caused by absorption and degradation of the energy. Sound of this intensity is difficult to be transmitted efficiently and as a means of raising the change insect behaviour, which differs remotes possibility in the controlling programme. Many insect cells are in the range of the ultrasonic. This is a sound of a frequency above about 20,000 cycles/second (below threshold to human ear). Infrasound (20 cycles) leads to setting up of resonance among the internal agents causing violent agitation and hemorrhage.

Centrifugal force for certain commodities can be used to away with the insect infestation. For example a mechanical device, called Entoleter at a speed of 2000-3000 rpm destroys insect infestation in milled products. However, this method is costly.

d. Host Resistance

With the advancement of science, a lot of crop varieties have been developed that show varying degree of resistance against different insect pests (Table 15.3). The resistance is due to the one or more biochemical, physiological, genetical and histological characters of the plant. Presently there are many varieties of cereals and pulses that show varying degree of resistance to insects.

Now these days, some genetically modified (transgenic) plants are available that carry genes for insect resistance. The most common example is pea having foreign gene (a -AI-Pv) resistant to pulse beetle and tobacco having α-ai gene resistant to *Tenebrio molitor*.

Table 15.3: Varieties of some crops resistant to stored grain insect pests

Insect	Crop and varieties	Reference
S. oryzae	**Sorghum** : CSV-15, Lakadi,	Reddy *et al.* (2002)
	Wheat : K-65, K-136, K-143	Tiwari & Sharma (2002)
R. dominica	**Wheat**: K-157, K-162, K-140	Tiwari & Sharma (2002)
	Rice: Cauvery, Bala, Jaya, IR-8	
	Barley: RS 6, RD 2627, K 719	Chander, 2003
C. chinensis	**Cowpea** : K-98, FTC-27, GC-2	Singh & Sharma (2003)
	Pea : RPG-3, BR-12, Rachna	Singh & Sharma (2003)
T. granarium	**Wheat:** Kalyansona, UP-301, WH-291	Jakhar *et al.* (2006)

e. Insect Growth Regulators

Alarm, dispersal, aggregation and sex attractant pheromones are known in a number of storage pests but studies have been most valuable with sex pheromones in almond moth, Indian meal moth and with aggregation pheromone of lesser grain borer. There are three possible application of pheromones to control storage pests, *viz.,* trapping with sex attractant, using aggregation pheromone with insect pathogen or chemosterilant and perfusing the warehouses atmosphere with pheromone to disrupt normal courtship behaviour of males. The compound 4,8-dimethyldecanal (or DMD) was found effective aggregation pheromone against *Tribolium* and compound sitophinone and grain volatiles (valeraldehyde, maltol, vanillin) were used in pheromonal control of *S. oryzae* (Phillips *et al.*, 1993). The adults of *T. castaneum* were also trapped by synthetic lures, *i.e,.* PT 6 Allure and Trappit (Mullen, 1992) under laboratory conditions. The 1-pentadecene and 1,6-pentadecadiene are also found effective aggregation pheromone for *Tribolium* (Amaud *et al.*, 2002). Ismoratou and Coworkers (2003) reported that synthesis of (4R)-methynnonan-1-ol and (4R, 8RS)-dimethyldeconal from (5)-(+)-3,7-dimethyl 1-1,6-octadiene were effective when 4R used as sex pheromone against yellow meal worm and 4R, 8RS used as aggregation pheromone against rust red flour beetle.

Juvenile hormone analogues : Ambika and Abraham (1982) reported that topical application of methoprene (JHA) at a dose of 1.0 and 0.5μ/4th instar larva gave 100 per cent mortality by 42 and 90 days after treatment, respectively. Such larval treatment resulted in the formation of supernumerary larva which died within 7-9 days. Bhargava and Devaraj Urs (1993) reported the ovicidal properties of IGRs, hydroprene, methoprene and fenoxycarb against *C. cephalonica*. Sharma and Bhargava (2000) observed that larval duration, length and width of larva, pupal weight and pupal period increased due to fenoxycarb treatment. Reduction in adult emergence increased with the increase in concentration of fenoxycarb.

Chapter 16

Multiple Choice Questions

Footnote: Bold means correct answer

Basic

1. Grain Stored Research and Training Center is situated ati.
   i) New Delhi ii) **Hapur**
   iii) Lucknow iv) Banglore
2. Central Food Technology Research Institute (CFTRI) is situated at
   i) **Mysore** ii) Hapur
   iii) Mumbai iv) Pune
3. The Department of food was established by the Government of India in the year:
   i) 1939 ii) **1942**
   iii) 1949 iv) 1952
4. Central food Technology Research Institute (CFTRI) is established in:
   i) 1945 ii) 1947
   iii) **1950** iv) 1954
5. The Central warehousing corporation was set up on:
   i) 2nd March, 1954 ii) **2nd March, 1957**
   iii) 2nd November, 1957 iv) 2nd December, 1957
6. The First state warehousing corporation was set up in ___________ in 1956.
   i) Rajasthan ii) UP
   iii) MP iv) **Bihar**

Birds

1. The head quarter of Salim all center for orthology and natural History is situated at:
   i) Bangluru ii) **Coimbatore**
   iii) New Delhi iv) Pune

2. The book " The Bird of india" a guide is Indian Orthology is written by:
   i) Salim Ali ii) Iswer Prakash
   iii) R.C Saxena iv) **Edwin H. Barnes**

3. Parrot Belongs to order:
   i) **Psittaciformes** ii) Columbiformes
   iii) Corividae iv) Passeriformers

4. Birds are Charactersized by the presence of:
   i) Homiothermic condition and viviparity
   ii) Horny jaws and four chambered heart
   iii) Ossified Skeleton and four chambered heart
   iv) **Bipedal Locomotion and body coverd feather**

5. Common name of *Acridotheres tristis* is
   i) House sparrow ii) House crow
   iii) **Myne** iv) Parrot

6. All India Co-ordinated Research Project on ornithology is situated at:
   i) Bangluru ii) Hyderabad
   iii) **New Delhi** iv) Lucknow

7. Birds can be clasified as:
   i) Insecticides ii) Graminivorus
   iii) Carnivorous iv) **All the above**

8. Common Myne is belongs to family
   i) **Saturnidae** ii) Columbidae
   iii) Ploceidae iv) Corvidae

9. The Scientific name of peacock is
   i) *Gallus gallus*　　ii) ***Paro cristatus***
   iii) *Columbialivia*　　iv) None of the above

10. The pneumatic bones are characteristic of
   i) Mammals　　ii) Repltiles
   iii) **Aves**　　iv) Lizards

11. What is the shidy of Bird eggs called?
   i) Ornithology　　ii) **Oology**
   iii) Ichthyology　　iv) Zoology

Rodents

1. Ishwar Prakash did work on:
   i) Termites　　ii) Mites
   iii) Birds　　iv) **Rats**

2. Tatera indica Belongs to sub-family
   i) Murinee　　ii) Gabillinee
   iii) Thripinee　　iv) Heliothripinee

3. Zinc Phosphide in used i poison baits
   i) 5%　　ii) **2%**
   iii) 3%　　iv) 1%

4. Which among the following is not anticoagulant:
   i) Warfarin　　ii) Chlorofacinone
   iii) Difacinone　　iv) **Chlolecalciferol**

5. The gravid female rat live in :
   i) Irregular burrow　　ii) Round burrow
   iii) **Semi-circular burrow**　　iv) None of the above

6. The scientific name of Indian mole rat :
   i) *Rattus rattus*　　ii) ***Bandicole bengalesis***
   iii) *Bandicola indica*　　iv) *Tetera Indica*

7. Prolin is mixture of :
   i) **Warfarin and Sulfaquinone**
   ii) Warfarin and Aluminium Phosphide
   iii) Warfarin and strychnin
   iv) Warfarin and Zink phosphide

8. The gestation period in rats is:
   i) **21-25 days** ii) 31-35 days
   iii) 41-45 days iv) 51-55 days

9. The Scientific name of brown spiny mouse is:
   i) *Golunda Ellioti* ii) ***Mus platythrix***
   iii) *Rattus exulans* iv) *Rattus rattus*

10. Warfarine is used as:
    i) Acaricide ii) Insecticide
    iii) Fungicide iv) **Rodenticide**

11. The life span of a rat is;
    i) **2-3 years** ii) 3-5 years
    iii) 6-8 years iv) 9-11 years

12. Rats begin to rapruduce at the age of :
    i) 1-2 months ii) **3-4 months**
    iii) 8-10 months iv) 10-12 months

13. The common name of *Rattus rattus* is:
    i) Large bandicoot rat ii) Soft furred rat
    iii) Field mouse iv) **House rat**

14. The Zinc Phosphide is the most widely used poison for rodent cetrol in india as:
    i) **Acute poison** ii) Chronic poison
    iii) Antiwagulant iv) Fumigent

15. The scientific name of five striped is:
    i) *Mrines hurrianee* ii) *Mus Boodaga*
    iii) ***Funambulus penenti*** iv) *Funambulus palmarum*

16. Silmurin is used as:
    i) Molluscacide ii) Acaricide
    iii) **Rudenticide** iv) Insecticide

17. Dark yellow colour on the label of container of pesticide indicates that pesticide is:
    i) Non-toxic ii) Moderately toxic
    iii) **Highly toxic** iv) Extremely toxic

18. The rats belong is phylum:
    i) Mollusca ii) Orthopoda
    iii) Chordala iv) Teterapoda

Storage Insect Pests

1. The Acid fuchsin method for determine the hidden infostation was described by:
    i) Pradhan (1948) ii) **Frankenfield (1948)**
    iii) Howe (1954) iv) Goosens (1949)

2. The determination of hidden infostation by Gentian violet method was suggested by
    i) Frankenfield (1948) ii) Milner *et al* (1950)
    iii) **Goosens (1949)** iv) Agarwal *et al* (1957)

3. Which of the following method for determination of hidden in infostation by Milner *et al.* (1950):
    i) **Berberine Sulphate stain**
    ii) Iodine solution method
    iii) Gention violet method
    iv) None of the above

4. Carbon dioxide determination for hidden infestation was described by
    i) Apt (1952) ii) White (1956)
    iii) Milner *et al.*(1950) iv) **Howe and oxley**

5. Angoumois gain moth belong to
   i) **Gelechiidae** ii) Dermestidae
   iii) Tenebrionidae iv) Curculionidae

6. The scientific name of case making cloth moth is
   i) *Tineola Bisselliella* ii) *Trichophega Tapetzella*
   iii) ***Tinea Pellionella*** iv) *Galleria Mellonella*

7. Which of the following insect infests the grain in the field and than comes to storage?
   i) *Lasioderma sericorne* ii) ***Sitotroga cereallela***
   iii) *Tribolium castaneum* iv) *Trogoderma granasium*

8. The dameging stage og Khapra beetle :
   i) **Grub** ii) Pupa
   iii) Adult iv) Grab and adult

9. All weevils belongs to family
   i) Anobiidae ii) Dermestidae
   iii) Tenebrionidae iv) **Curcutionidae**

10. Internal feeder is :
    i) *Tribolium Cartaneum* ii) *Corcyra Cephalonica*
    iii) *Trogoderma granarium* iv) ***Sitophilus oryzae***

11. The Infestation at one pest continues at zero percent moisture it is
    i) Lesser grain borer ii) Grain moth
    iii) **Khapra beetle** iv) Pulse beetle

12. Potato Tuber moth caused damege at
    i) Field ii) Storage
    iii) **Both field and storage** iv) None of these

13. The book entitled "Agricultural Pests of south Asia and Management" is written by
    i) David, B.V ii) **Atwal, A. S and Dhaliwal, G.S**
    iii) Bhargava, M.C and Kumawat, K.C iv) Srivastava, K.P

14. *Corcyra Cepholonica* belongs to order:
    i) Coleoptera ii) Diptera
    iii) **Lepidoptera** iv) Orthotera

15. Directorate of plant protection and Quarantine is situated at
    i) **Faridabad** ii) FaridKot
    iii) Lucknow iv) New Delhi

16. The Scientific name of drug store beatle is:
    i) *Lasioderme serricorne*
    ii) ***Stegobium Paniceum***
    iii) *Callosobruchus chinensis*
    iv) *Sitophilus oryzae*

17. Which one of the following stored grain beetle produces lot of powedary material when it attack grain:
    i) ***Rhizopertta dominica***
    ii) *Callosobruches chinensis*
    iii) *Sitophilus oryzae*
    iv) *Trogoderma granarium*

18. Natrional research center for interdrated Pest management is situated in
    i) Lucknow ii) Banglore
    iii) **New Delhi** iv) Jaipur

19. The Fumigation of rat burrows with
    i) EDB ii) Methyl bromide
    iii) Atropine iv) **Aluminium Phosphide**

15. Lesser Grain borer is belongs is family:
    i) **Bostrychidae** ii) Anthribidae
    iii) Anobiidae iv) Darmestidae

16. The scientific name of pulse beetle is:
    i) *Lasioderma serricorne*
    ii) ***Callosobruchus Chinensis***
    iii) *Attagenus piceus*
    iv) *Oryzaephilus Surinamensis*

17. Black Carpet beetle belongs to family:
    i) **Dermestidae** ii) Cucujidae
    iii) Bruchudae iv) Anthiribidae

18 X-ray method for detection of internal Infestation was described by:
    i) Dennis and Decker ii) **Milner**
    iii) White iv) How and Oxley

19. The total no. of eggs laid by khapra beatle is
    i) **100 - 150 eggs** ii) 200 - 250 eggs
    iii) 300-350 eggs iv) 400 - 450 eggs

20. The dosage of EDB and MB mixture for oil seed is
    i) **1.5 - 2.0 lb/ 1000 m^3** ii) 2.0 - 3.0 lb/ 1000 m^3
    iii) 3.0 - 4.0 lb/ 1000 m^3 iv) 4.0 - 5.0 lb/ 1000 m^3

21. Hydrogen cynide (HCN) is available in the form of:
    i) Bromofume ii) Durofume
    iii) Phosfume iv) **Cynogae**

22. Which of the following pest belong to family Gelechiidae
    i) *Galleria mellonella* ii) ***Sitotroge Cerelella***
    iii) *Cydia Pomonella* iv) *Corcyra Cephalonica*

23. Grain-O-Cide fumigant is a mixture of:
    i) Methyl Bromide and Carbon tetrachloride
    ii) **Carbon disulphide and carbon tetrachloride**
    iii) Ethylene disbromide and methyl bromide
    iv) Ethylene dibromide and Carbon tetrachloride

24. Which of the following is not a lepidopterous pest
    i) *Corcyra cephalonica*
    ii) *Ephestia cautella*
    iii) *Lasioderma serricorne*
    iv) *Plodia interpunctella*

25. Quantity of alumium oxide in aluminum phosphide is :
    i) 55 % ii) 40%
    iii) 20 % iv) **5 %**

26. The scientific name of coffee bean weevil is
    i) *Sitophilus oryzae* ii) ***Araecerus fesciculatus***
    iii) *Sitophilus grannarius* iv) *Lasioderma serricorne*

27. Ninhydrine methode for detection of internal infestation was described by
    i) **Dennis and Decker** ii) Milner
    iii) White iv) Howe and Oxley

28. Which of the following is a major pest of stored pulses
    i) *Rhizopertha dominica*
    ii) *Trogoderma Granasium*
    iii) ***Callosobruchus chinensis***
    iv) *Tribolium castaneum*

29. Flat grain beetle belong is family:
    i) Ptinidae ii) **Cucujidae**
    iii) Tenebrionidae iv) Bruchidae

30. The scientific name of stored nut moth is
    i) ***Aphomia gulares*** ii) *Cadra cautella*
    iii) *Plodia interpunctella* iv) *Tinae pellionalla*

31. The total Number of eggs laid by cigarette beetle is
    i) **100-150 eggs** ii) 200-350 eggs
    iii) 400-550 eggs iv) 600-750 eggs

32. The Pusa bin was designed by:
    i) Milner (1950)
    ii) **Pradhan and Mukhuejee (1968)**
    iii) Bhargava and Kumawat (1995)
    iv) Dennis and Decker (1964)

33. The scientific name of grain mite is
    i) *Tyrophagus casei* ii) *Glycyphogus domesticus*
    iii) *Carpoglyphus loctis* iv) ***Acarus siro***

34. Insect Probe trap device was developed by
    i) PAU ii) CFTRI
    iii) **TNAU** iv) SKRAU

35. The book "Stored Grain Pests and Their Management" is authored by
    (A) **BP Khare** (B) DW Hall
    (C) Subramanyam B & Hagstrum DW (D) V Khader

36. The author of "*Handling and Storage of Food Grains in Tropical and Subtropical Areas*" is
    (A) BP Khare (B) **DW Hall**
    (C) Subramanyam B & Hagstrum DW (D) V Khader

37. The author of "*Interrelated Management of Insects in Stored Products*" is
    (A) BP Khare (B) DW Hall
    (C) **Subramanyam B & Hagstrum DW** (D) V Khader

38. The estimated losses in stored food grains as per FAO is
    (A) 5-8% (B) **8-10%**
    (C) 10-15% (D) 15-20%

39. The optimum temperature for growth and development of stored pests is
    (A) 18-28°C (B) 22-28°C
    (C) **28-32°C** (D) 30-35°C

40. The insect pests associated with hidden infestation is
    (A) Paddy moth (B) Pulse beetle
    (C) **Both A and B** (D) None of the above

41. The infestation of fresh stock from the pest already presented in the godown is
    (A) Hidden infestation (B) **Cross infestation**
    (C) Both A and B (D) None of the above

42. The physical method for detection of infestation of stored pest
(A) NMR and IR spectroscopy (B) Berleese method
(C) Floatation method (D) **All of the above**

43. The Chemical method for detection of infestation of stored pest
(A) ELISA test (B) Uric acid anaysis
(C) Specific gravity method (D) **All of the above**

44. The staining technique for detection of infestation of stored **pest**
(A) Ninhydrin method (B) Egg plugs
(C) **Both A and B** (D) None of the above

45. The physical method for detection of infestation of stored pest
(A) NMR and IR spectroscopy (B) Berleese method
(C) Floatation method (D) **All of the above**

46. The presence of exit holes in whole spices indicate the infestation of
(A) *Ephestia cautella* (B) ***Stegobium paniceum***
(C) *Sitophilus oryzae* (D) *Callosobruchus*sp**.**

47. The presence of pupal cases sticking to peanut shells and gunny bags indicate the infestation of
(A) *Ephestia cautella* (B) *Stegobium paniceum*
(C) Sitophilus oryzae (D) Caryedon serratus

48. The stored pests which attack sound grains are known as
(A) Secondary pest (B) **Primary pest**
(C) Tertiary pest (D) None of the above

49. The stored pests which attack broken grains are known as
    (A) **Secondary pest** (B) Primary pest
    (C) Detritus pest (D) None of the above

50. The stored pests which attack unhealthy grains are known as
    (A) Secondary pest (B) Primary pest
    (C) **Detritus pest** (D) None of the above

51. Which of the following is a detritus stored pest
    (A) *Ephestia cautella* (B) *Sitotroga cerelella*
    (C) *Sitophilus oryzae* (D) **Corn sap beetle**

52. Which of the following is a primary stored pest
    (A) *Ephestia cautella* (B) *Sitotroga cerelella*
    (C) *Sitophilus oryzae* (D) **All of the above**

53. Which of the following is a primary stored pest
    (A) *Rhizopertha dominica* (B) *Callosobruchus* sp.
    (C) *Corcyra cephalonica* (D) **All of the above**

54. Which of the following is a secondary stored pest
    (A) *Tribolium castaneum* (B) *Cryptolestes ferrugineus*
    (C) *Oryzaephilus surinamensis* (D) **All of the above**

55. The rice weevil can be identified by
    (A) Having snout/beak (B) Antennae Geniculate
    (C) Four reddish or yellowish spot on elytra (D) **All of the above**

56. The rice weevil can be identified by
    (A) Having snout/beak (B) Antennae Geniculate
    (C) Four reddish or yellowish spot on elytra (D) **All of the above**

57. The typical damaging symptom of rice weevil is
    (A) **Circular exit hole** (B) Irregular exit hole
    (C) Split grains (D) None of the above

58. The typical damaging symptom of Lesser grain borer is
    (A) Circular exit hole (B) **Irregular exit hole**
    (C) Split grains (D) None of the above

59. The damaging stage of rice weevil is
    (A) Adultonly (B) Larva only
    (C) **Both Larva and adults** (D) None of the above

60. The Lesser grain borer can be identified by
    (A) Head deflexed downward
    (B) prominent constriction between prothorax and elytra
    (C) antenna 10 segmented ending in tripartite club
    (D) **All of the above**

61. The damaging stage of Lesser grain borer is
    (A) Adultonly (B) Larva only
    (C) **Both Larva and adults**(D) None of the above

62. The Family of Lesser grain borer is
    (A) **Bostrichidae** (B) Tenebrionidae
    (C) Dermestidae (D) Bruchidae

63. The Family of Khapra beetle is
    (A) Bostrichidae (B) Tenebrionidae
    (C) **Dermestidae** (D) Bruchidae

64. The Family of Red flour beetle is
    (A) Bostrichidae (B) **Tenebrionidae**
    (C)Dermestidae (D) Bruchidae

65. The scientific name of Khapra beetle is
(A) *Rhizopertha dominica* (B) ***Trogoderma granarium***
(C) *Tribolium csataneum* (D) *Sitophilus oryzae*

66. The scientific name of Lesser grain borer is
(A) ***Rhizopertha dominica*** (B) *Trogoderma granarium*
(C) *Tribolium csataneum* (D) *Sitophilus oryzae*

67. The damaging stage of Khapra beetle is
(A) Adultonly (B) **Larvaonly**
(C) Both Larva and adults (D) None of the above

68. The hibernating stage of Khapra beetle is
(A) Adultonly (B) **Larvaonly**
(C) Both Larva and adults (D) None of the above

69. The Larvae of Khapra beetle is
(A) **Positively thigmotactic**(B) Negatively thigmotactic
(C) Thigmotactic neutral (D) None of the above

70. Which of these is a strong flier causing cross infestation
(A) **Lesser grain borer** (B) Khapra beetle
(C) Rice weevil (D) Rice weevil

71. The stored pest capable of surviving under low moisture (<4%) of seed
(A) Lesser grain borer (B) **Khapra beetle**
(C) Rice weevil (D) Rice weevil

72. Which of these is a strong flier causing cross infestation
(A) **Lesser grain borer** (B) Khapra beetle
(C) Red rust flour beetle (D) Rice weevil

73. The Khapra beetle can be identified by
    (A) Adults having no distinction of head, thorax and abdomen
    (B) Antennae clubbed
    (C) Larva have tuft forming tail at posterior end
    (D) **All of the above**

74. The Oxygen requirement for survival of Khapra beetle eggs **is**
    (A) 19.8% (B) 18.8%
    (C) **16.8%** (D) 14.8%

75. Which of these shows cannibalism
    (A) Lesser grain borer (B) Khapra beetle
    (C) **Red rust flour beetle** (D) Rice weevil

76. Which of Tribolium species is a flyer
    (A) ***Tribolium castaneum*** (B) *Tribolium confusum*
    (C) Both A and B (D) None of the above

77. The damaging stage of Red rust flour beetle is
    (A) Adult only (B) Larva only
    (C) **Both Larva and adults**(D) None of the above

78. The offensive pungent smell of infested grain due to Red rust flour beetle is because of
    (A) **Gaseous quinones** (B) Gaseous phenols
    (C) Gaseous acids (D) None of the above

79. The Red rust flour beetle can be identified by
    (A) All three body segments are distinctly visible dorsally
    (B) Antennae clubbed
    (C) Larva negatively phototactic
    (D) **All of the above**

80. The hibernating stage of Pulse beetle is
    (A) Adultonly (B) **Larvaonly**
    (C) Both Larva and adults (D) None of the above

81. Which of the following is both field and storage pest
    (A) *Chromatomyia horticola* (B) *Lampides boeticus*
    (C) *Ophiomyia phaseoli* (D) ***Callosobruchus chinensis***

82. The Pulse beetle can be identified by
    (A) Spine on each of the inner and outer edges of the of hind femur
    (B) Antennae Serrate
    (C) Elytra truncate not covering the pygidium
    (D) **All of the above**

83. The scientific name of the spotted bean bruchid is
    (A) *Callosobruchus chinensis*
    (B) ***Callosobruchus maculatus***
    (C) *Callosobruchus phaseoli*
    (D) *Callosobruchus analis*

84. The scientific name of the Lentil bruchid/Dhora is
    (A) *Callosobruchus chinensis* (B) *Callosobruchus maculatus*
    (C) ***Callosobruchus phaseoli*** (D) *Callosobruchus analis*

85. The scientific name of the Moong bruchid/dhora is
    (A) *Callosobruchus chinensis* (B) *Callosobruchus maculatus*
    (C) *Callosobruchus phaseoli* (D) ***Callosobruchus analis***

86. The scientific name of the Pea bruchid/dhora is
    (A) ***Callosobruchus affinis*** (B) *Callosobruchus maculatus*
    (C) *Callosobruchus phaseoli* (D) *Callosobruchus analis*

87. The typical damaging symptom of Pulse beetle is
    (A) Circular exit hole (B) Presence of chambers in seed
    (C) **Both A and B** (D) None of the above

88. The Family of Cigarette beetle is
    (A) **Anobiidae** (B) Tenebrionidae
    (C) Phycitidae (D) Silvanidae

89. The Family of Saw tooth grain beetle is
    (A) Anobiidae (B) Tenebrionidae
    (C) Phycitidae (D) **Silvanidae**

90. The Family of Black fungus beetle is
    (A) Anobiidae (B) **Tenebrionidae**
    (C) Phycitidae (D) Silvanidae

91. The Family of Almond moth is
    (A) Anobiidae (B) Tenebrionidae
    (C) **Phycitidae** (D) Silvanidae

92. The scientific name of Cigarette beetle is
    (A) ***Lasioderma serricorne*** (B) *Stegobium panaceum*
    (C) *Oryzaephilus surinamensis* (D) *Ephestia cautella*

93. The scientific name of Drug store beetle is
    (A) *Lasioderma serricorne* (B) ***Stegobium panaceum***
    (C) *Oryzaephilus surinamensis* (D) *Ephestia cautella*

94. The scientific name of Almond moth is
    (A) *Lasioderma serricorne* (B) *Stegobium panaceum*
    (C) *Oryzaephilus surinamensis* (D) ***Ephestia cautella***

95. The scientific name of Saw tooth grain beetle is
    (A) *Lasioderma serricorne* (B) *Stegobium panaceum*
    (C) ***Oryzaephilus surinamensis*** (D) *Ephestia cautella*

96. The Family of Angoumois grain moth is
    (A) **Gelechiidae** (B) Pyralidae
    (C) Phycitidae (D) Pyraustidae

97. The Family of Indian meal moth is
    (A) Gelechiidae (B) **Pyralidae**
    (C) Phycitidae (D) Pyraustidae

98. The Family of Rice moth is
    (A) Gelechiidae (B) **Pyralidae**
    (C) Phycitidae (D) Pyraustidae

99. The scientific name of Indian meal moth is
    (A) *Sitotroga cerealella* (B) *Corcyra cephalonica*
    (C) ***Plodia interpunctella*** (D) *Ephestia cautella*

100. The scientific name of Rice moth is
    (A) *Sitotroga cerealella* (B) ***Corcyra cephalonica***
    (C) *Plodia interpunctella* (D) *Ephestia cautella*

101. The scientific name of Angoumois grain moth is
    (A) ***Sitotroga cerealella*** (B) *orcyra cephalonica*
    (C) *Plodia interpunctella* (D) *Ephestia cautella*

102. The Sitotroga cerealella can be identified by
   (A) Narrow pointed wings with long hairs along posterior margin
   (B) Circular exit hole with a flap/trap door
   (C) Larva white with yellow brown head
   (D) **All of the above**

103. The hibernating stage of Sitotroga cerealella is
   (A) Adult only (B) **Larva only**
   (C) Both Larva and adults (D) None of the above

104. The webbing of grains indicate the infestation of
   (A) *Plodia interpunctella* (B) *Corcyra cephalonica*
   (C) **Both A and B** (D) None of the above

105. The silk tunnelling in the grains indicate the infestation of
   (A) *Plodia interpunctella* (B) *Corcyra cephalonica*
   (C) ***Ephestia cautella*** (D) None of the above

106. Clogging of mill machinery due very high level of silk tunnels in the grains is due to the infestation of
   (A) *Plodia interpunctella* (B) *Corcyra cephalonica*
   (C) ***Ephestia cautella*** (D) None of the above

107. The amount of food eaten by Rodents in a single day is
   (A) **10% of body weight** (B) 15% of body weight
   (C) 05% of body weight (D) 01% of body weight

108. The amount of food damaged by Rodents in a single day is
   (A) 10 times the food they consume
   (B) **20 times the food they consume**

(C) 30 times the food they consume

(D) None of the above

109. The causal organism of Rat bite fever is

(A) *Streptobacillus moniliformis* (B) *Spirillum minor*

(C) **Both A and B** (D) None of the above

110. Thescientific name of Black rat/Roof rat/House rat is

(A) ***Rattus rattus*** (B) *Mus musculus*

(C) *Bandicota bengalensis* (D) *Bandicota indica*

111. The scientific name of House mouse is

(A) *Rattus rattus* (B) ***Mus musculus***

(C) *Bandicota bengalensis* (D) *Bandicota indica*

112. The scientific name of smaller bandicot is

(A) *Rattus rattus* (B) *Mus musculus*

(C) ***Bandicota bengalensis***(D) *Bandicota indica*

113. The scientific name of larger bandicot is

(A) *Rattus rattus* (B) *Mus musculus*

(C) *Bandicota bengalensis* (D) ***Bandicota indica***

114 The scientific name of Ship rat or Brown rat is

(A) *Rattus rattus* (B) *Mus musculus*

(C) *Bandicota bengalensis* (D) ***Rattus norvegicus***

115. The scientific name of Gebril is

(A) ***Tatera indica*** (B) *Mus platythrix*

(C) *Hystrix indica* (D) *Funambulus palmarum*

116. The Chemosteriliants for rodent control is

(A) Tepa (B) Thiotepa

(C) Apholate (D) **All of the above**

117. The scientific name of Field mice is
   (A) *Tatera indica* (B) ***Mus platythrix***
   (C) *Hystrix indica* (D) *Funambulus palmarum*

118. The scientific name of Squirrel is
   (A) ***Tatera indica*** (B) *Mus platythrix*
   (C) *Hystrix indica* (D) ***Funambulus palmarum***

119. The scientific name of Procupines is
   (A) *Tatera indica* (B) *Mus platythrix*
   (C) ***Hystrixindica*** (D) *Funambulus palmarum*

120. The Primary vector of Plague caused by Yersinia pestis, a bacterium is
   (A) *Rattus rattus* (B) *R. norvegicus*
   (C) *B. bengalensis* (D) ***Xenopsylla cheopis***

121. The Secondary vector of Plague caused by Yersinia pestis, a bacterium is
   (A) *Rattus rattus* (B) *R. norvegicus*
   (C) *B. bengalensis* (D) **All of the above**

122. Sweating of food grains is more under which storage conditions
   (A) **Bulk storage** (B) Bag storage
   (C) Air tight storage (D) All of the above

123. Heating of food grains is more under which storage conditions
   (A) **Underground storage** (B) Bag storage
   (C) Air tight storage (D) All of the above

124. Indian Grain Storage Institute is located at
   (A) Faridabad (B) **Hapur**
   (C) Ludhiana (D) New Delhi

125. Indian Grain Storage Institute (earlier GSRTC) was established in the year

(A) **1958** (B) 1968

(C) 1976 (D) 1988

126. Pusa bin is

(A) Outdoor bin (B) **Indoor bin**

(C) Both A and B (D) None of the above

127. Which of the following is the best storage structure

(A) Gharelu Theka (B) Pusa bin

(C) Ferro-cement bin (D) **Vaccum Process storage**

128. Which of the following is Bamboo outdoor storage structure

(A) Galgai (B) Bukari

(C) Topa (D) **All of the above**

129. Which of the following is Timber made outdoor storage structure

(A) Gutibharal (B) Peti

(C) Pathyam (D) **All of the above**

130. Which of the following is Paddy straw made outdoor storage structure

(A) Pura (B) Puri

(C) Morai (D) **All of the above**

131. Which of the following is Mud/brick made outdoor storage structure

(A) Bharola (B) Pusa Bin

(C) Kothi (D) **All of the above**

132. Which of the following is under ground storage structure

(A) Peve (B) Khatti

(C) Bandha (D) **All of the above**

133. The optimum moisture content of food grains for storage is
(A) 5% (B) **10%**
(C) 15% (D) None of the above

134. The concentration of O_2 and CO_2 in Controlled Atmosphere Storage (CAS) is
(A) **2-4% O_2 and 9-9.50% CO_2**
(B) 10-14% $O_2$and 5- 8% CO_2
(C) 6-10% $O_2$and 4-8% CO_2
(D) None of the above

135. The insecticides recommended as prophylactic measure in storage is
(A) Malathion (0.50%)
(B) Pyrethrum with 2%pyrethrin
(C) Primiphos methyl (0.50%)
(D) **All of the above**

136. The application rate of Malathion (0.50%) recommended as prophylactic measure in storage is
(A) 1 L/100m^3 (B) 2 L/100m^3
(C) **3 L/100m^3** (D) 5 L/100m^3

137. The application rate of Aluminium Phosphide in storage **is**
(A) 2 tablet/ton (B) **25 tablets/100m^3**
(C) 3 tablet/ton (D) 50 tablets/100m^3

138. The weight of tablet of Aluminium Phosphide in storage **is**
(A) 2 g (B) **3 g**
(C) 5 g (D) 10 g

139. The toxicity of Aluminium Phosphide is due to the
(A) Phostoxin gas (B) **Phosphine gas**
(C)Celphos gas (D) All of the above

140. The Aluminium Phosphide is sold under the trade name of
(A) Phostoxin (B) Delicia
(C) Celphos (D) **All of the above**

141. As per Insecticide Act 1968, the use of Aluminium Phosphide is under
(A) Open use category (B) **Restricted use category**
(C) Banned category (D) All of the above

142. Each Tablet ofCelphos consists of
(A) **AlP (56%) + carbamate (44%)**
(B) AlP (50%) + carbamate (50%)
(C) AlP (60%) + carbamate
(D) AlP (44%) + (40%)carbamate (56%)

143. The application rate of Ethylene Dibromide (EDB) in storage is
(A) **3 ml/q** (B) 5 ml/q
(C) 6 ml/q (D) None of the above

144. The which of the following is highly inflammable
(A) **Carbon tetrachloride** (B) EDB
(C) HCN (D) ALP

145. The proportion of Carbon tetrachloride mixed with Dichloro ethane is
(A) **1:3** (B) 1:5
(C) 1:10 (D) 1:20

146. The Carbon tetrachloride is mixed with Dichloro ethane to
(A) To increase shelf life (B) **Reduce fire hazard**
(C) To increase toxicity (D) To reduce toxicity

147. The acute poison which kill rats by exposure to poison bait is known as
(A) Multiple dose poison (B) **Single dose poison**
(C) Both A and (D) None of the above

148. The chronic poison which kill rats by repetitive exposure to poison bait is known as
   (A) **Multiple dose poison** (B) Single dose poison
   (C) Both A and B (D) None of the above

149. Which of the following is single dose poison
   (A) ANTU (B) Zinc Phosphide
   (C) Norbromide (D) **All of the above**

150. Which of the following is single dose poison
   (A) Strychnine (B) Arsenous trioxide
   (C) Barium Carbonate (D) **All of the above**

151. Which of the following is second generation single dose poison
   (A) Bromadiolone (B) Vacour
   (C) Silmurin (D) **All of the above**

152. Brodifacoum is a
   (A) Anticoagulant (B) Single dose poison
   (C) second generation acute poison (D) **All of the above**

153. Which of the following is multiple dose poison
   (A) Warfarin (B) Rodofarin
   (C) Racumin (D) **All of the above**

154. Which of the following is multiple dose poison
   (A) Tomafarin (B) Ratafin
   (C) Racumin (D) **All of the above**

155. Bromadiolone, a single dose anticoagulant is sold under the trade name of
   (A) Talon (B) **Makki**
   (C) RH-787 (D) None of the above

156. Brodifacoum, a single dose anticoagulant is sold under the trade name of
(A) **Talon** (B) Makki
(C) RH-787 (D) None of the above

157. Vacour, a single dose anticoagulant is sold under the trade name of
(A) Talon (B) Makki
(C) **RH-787** (D) None of the above

158. Which of the following cause Neophobia in rodents
(A) Multiple dose poison (B) **Single dose poison**
(C) Both A and (D) None of the above

159. Rodent control should be organised on
(A) Individual basis (B) **Community basis**
(C) Both A and (D) None of the above

160. The Botanical used for Rodent control
(A) **Sweet flag** (B) Sweet potato
(C) Yam (D) None of the above

161. The Sulfuryl fluoride, a fumigant is sold under the trade name of
(A) Vapourmate (B) **Profume**
(C) Ecofume (D) None of the above

162. The Ethyl formate, a fumigant is sold under the trade name of
(A) **Vapourmate** (B) Profume
(C) Ecofume (D) None of the above

163. The mixture of fumigant Phosphine @% + CO_2 98% is sold under the trade name of
(A) Vapourmate (B) Profume
(C) **Ecofume** (D) None of the above

164. Calcium Cyanide, a fumigant for rodent control is sold under the trade name of

(A) **Cyanogas** (B) Raticide

(C) Celphos (D) None of the above

165. Nor Bromide, a fumigant for rodent control is sold under the trade name of

(A) Cyanogas (B) **Raticide**

(C) Celphos (D) None of the above

166. The Chemosteriliants for rodent control is

(A) Colchicine (B) Furadantine

(C) Hempa (D) **All of the above**

References

Agrawal, N.S. and Christensen, C.M. and Hodson, A.C., 1957. Grain storage fungi associated with the granary weevil. *J. econ. Ent.*, 50: 312-315

Amaud, L.; Lognay, G.; Verscheure, M.; Leenaers, L.; Gaspar, C. and Haubrug, E. 2002. Is dimethyldeconal a common aggregation pheromone of *Tribolium* flour beetle. *J. Chem. ecol.* 28 (3) : 523-532.

Ambika, B. and Abraham, C.C. 1982. Effect of juvenile hormone analogue, methoprene (ZR-515) on development of eggs and larvae of *Corcyra cephalonica* Stainton (Lepidoptera : Pyralidae). *Kerla Agril. Res. J.*, 20: 60-62.

Apt, A.C. 1950. A method of detecting hidden infestation in wheat. *Milling production.* 15 (5): 1.

Apt, A.C. 1952. A rapid method of examining wheat samples for infestation. *Milling production.* 17 (5): 4.

Awaknavar, J.S.; Krishnamurthy, K.C.; Babu, C.K. and Ramakumar, M.V. 1989. Different storage structures for safe storage of pulses. *Mysore J. Agri. Sci.*, 23 : 526- 532

Bhargava, M.C. and Devaraj Urs, KC. 1993. Ovicidal effect of three insect growth regulators on *Corcyra cephalonica. Indian J. Pl. Prot.* 21 (2) : 195-197.

Chander, R. 2003. Host preference and bio-control studies of *Rhizopertha dominica* (Fab.) on barley and its management. Ph.D. (Thesis) submitted to Rajasthan Agricultural University, Bikaner, pp. 224.

Crook, L.J. 1962. The susceptibility of the red-rust flour beetle *Tribolium castaneum* (Herbst.) to gamma radiation. AERE-3889, *United Kingdom Atomic Energy Authority, Research Crop, Isotope Research Div*, Wantage, Berks, England. pp.-19.

Davich, T.B. and Lindquist, D.A. 1962. Exploratory studies on Gamma radiation for the sterilization of the boll weevil. *J. Econ. Ent.* 55 (2) : 164-167

Dean, G.A. 1911. Heat as a means of controlling mill insects. *J. Econ. Entomol.*, 4 : 142-158.

Dennis, N.M. and Decker, R. W.1962. A method and machine for detecting living internal insect infestation in wheat. *J.Econ. Ent.* 55 (2): 199-203.

Dick, K.M. 1987. Pest Management in stored groundnuts. *ICRISAT Information Bulletin* No. 22, Patancheru, AP. pp 28.

Doharey, R.B.; Srivastava, P.K. and Giris, G.K. 1975. Studies on the assessment of losses in wheat in Punjab. *Bull. Grain Tech.*, 13 (13): 159-161.

Dortbudak, N. 1975. Studies on the biology, ecology and methods of control of *Bruchus eryii* F. on lentil around Mardin province. *Rew. Appl. Ent.*, 65 : 738, 2740.

Erdman, H.E. 1962. X-rays effects on single and mixed species population of *Tribolium confusum* and *Tribolium castaneum* (Coleoptera : Tenebrionidae). HW-SA-2686, *General Electric Co. Hanford Atomic Products Operation*, Richland, Wash. pp. 8.

Fletcher, T.D. 1914. *Some South Indian Insects.* Superintendant of Government Printing Press, Madras, pp 565.

Frankenfield, J.C. 1948. Staining methods for dectecting weevil infestation in grain. *U.S.D.A. Burl. Ent and Pl. Quarantine Circ. ET-256.* pp. 4. Mimeographed

Girish, G.K. and Jain, S.K. 1974. Studies on efficacy of neem kernel powder against stored grain pests. *Bull. Grain Tech.*, 12: 226-228.

Harris, K.L.; Knudsen, L.F.; Nicholson, J.F.; Randolf, L.K. and Trawick, J.L. 1952. An investigation of insect and rodent contamination of wheat and wheat flour. *J. Assoc. Offic. Agric. Chemists.* 35 (1): 115-158.

Harvey, J.M. 1963. Irradiation of fruits and vegetables in a mobile cobalt 60 unit. *Division of Isotopes Development and Division of Biology and Medicine*, AEC. pp 55-57

Hoew,R.W. and Oxley,T.A. 1944. The use of carbon dioxide production as a measure of infestation in grain by insects. *Bull. Ent. Res.* 35: 11-22.

Hoover, D.L.; Floyd, E.H. and Richardson, H.D. 1963. Effects of 300 KV X-ray radiation on *Sitophilus oryzae. J. Econ. Ent.* 56 (5) : 584-586.

Kansu, A. 1962. Preliminary experiment of the sterilization of the pupae of the khapra beetle by irradiation with Gamma rays. *Z. Angew Ent.* 42 (2): 224-228.

Kauraw, L.P. and Prakash, A. 1980. Influence of storage fungi on the development of lesser grain borer in paddy cultivar CR-1084. *Seed Research* 8(2): 67-70

Ismuratov, G.Yu; Yakovelva, M.P.; Galyautdinova, A.V. and Tolstikov, G.A. 2003. Synthesis of (4R)-methynnonan-1-ol and (4R, 8RS)-dimethyldeconal from (5)-(+)-3,7-dimethyl 1-1,6-octadiene. *Chemistry of Natural Compounds.* 39 (1) : 31-33.

Jotwani, M.G. and Sircar, P. 1965. Neem seed as protectant against stored grain pests infesting wheat seed. *Indian J. Ent.* 27: 160-164.

Lal, S. 1984. Assessment of insect damage in stored wheat at farm level and studies on the impact of introduced new storage technology. Ph. D. Thesis submitted to Bhopal University, Bhopal, pp.138.

Lefroy, H.M. 1906. *Indian Insect Pests.* Superintendant of Government Printing Press, India, Calcutta pp 313.

MacLeay, W. 1837. In Dejean, P.F. Catalogue des Coleopteres de la Collection, *Troisieme Edition, Revue, Corrigee et Augmenfee Paris* XIV: 503 pp.

Meena, B.L. and Bhargava, M.C. 2003a. Effect of different storage containers on the incidence of *Corcyra cephalonica* in stored kernels of groundnut. *Indian J. Appl. Ent.*, 17 : 41.43.

Milner,M.; Barney, D.L. and Shellenberger, J.A. 1950. Use of selective fluorescent stains to detect insect egg plugs on grain kernels. *Science*, 12: 791-792.

Mookherjee, P.B.; Jotwani, M.G.; Yadav, T.D. and Sircar, P. 1968. Disinfestation of stored seeds by heat treatment. *Indian J. Ent.,* 30 : 197-202.

Mullen, M.A. 1992. Development of a pheromone trap for monitoring *Tribolium castaneum*. *J. Stored Prod. Res.* 28 (4) : 245-249.

Narsimhan, K.S.; Gawal, M.H. and Doharey, R.B., 1985. Impact of containers on seed paddy. *Bull Grain Tech*., 23(3): 269-277

Pelerents, C. 1963. Quelques resultants supplementaries Concernant L'influence Des Rayons Gamma Sor Les Chrysalides Et Les Ceufs De La Teigne De La Farine '*Ephestia kuehniella* Z.' *Reu. Agric.* 16 (2) : 305-320.

Pendlebury, J.B.; Jefferies, D.J.; Banham, E.J. and Bull, J.O. 1962. Some effects of Gamma radiation on the lesser grain borer (*Rhizopertha dominica* F.), tropical warehouse moth (*Cadra (Ephestia) cautella* Wlk.), Indian meal moth (*Plodia interpunctella* Hubn.) and the Cigarette beetle (*Lasioderma serricorne* F.). AERE-R-4003, United Kingdom Atomic Energy Authority. Research Groups. Isotope Research Div. Wantage, Berks, England, p. 23.

Pesson, P. 1963. Some experimental data on cobalt 60 radiation doses capable of arresting insect infestation of cereal and flour. *Food Irrad.* 3 (4) :A8-A21.

Phillips, T.W.; Jiang, X.L.; Burkholder, W.E.; Phillips, J.K. and Tran, H.Q. 1993. Behavioral responses to food volatiles by two species of stored product Coleoptera, *Sitophilus oryzae* (Curculionidae) and *Tribolium castaneum* (Tenebrionidae). *J. Chem. Ecol.* 19 (1): 23-734.

Pradhan, S. and Mookherji, P.B. 1969. Indian Council of Agricultural Research. *Tech. Bull.* (Agric.) 21: 1-10.

Prakash, A. and Kauraw, L.P., 1982b. Problems and quality control in paddy seed. *Seed and Farms,* 8(11): 13-16

Prakash, A.; Rao, J.; Pasalu, I.C. and Mathur, K.C., 1987. Rice storage and insect pest management. B.R. Publishing Corporation, New Delhi, pp. 337

Quraishi, M.S. and Metin, M. 1963. Radiosensitivity of various stages of *Callosobruchus chinensis* L. In "Radiation and radioisotopes applied to insects of Agricultural importance. Proceeding of symposium, Athens, 22-26 April, Vienna, International Atomic Energy Agency. pp 479-484.

Reddy, K.P.K.; Reddy, K.D. and Singh, B.U. 2002. Components of resistance in sorghum to the rice weevil, *Sitophilus oryzae*, Antixenosis. *I. J. Ent.,* 64 (3): 262-274.

Saxena, B.P.; Sharma, P.R.; Thappa, R.K. and Tikku, K. 1992. Temperature induced sterilization for control of three stored grain beetles. *J. Stored Prod. Res.*, 28 : 67-70.

Senguttuvan, T.; Kareem, A.A. and Rajendran, R. 1993. Effect of different storage containers in storing groundnut against rice moth, *Corcyra cephalonica* Stainton (Galleriidae : Lepidoptera). *Bull. Grain Tech.*, 31: 129-131.

Sharma, K.C. and Bhargava, M.C. 2000. Effect of juvenile hormone analogue (fenoxycarb) on rice moth, *Corcyra cephalonica* Staint. (Lepidoptera : Pyralidae). *Bull. Grain Tech.* 33 (1) : 59-65.

Sharma, P.R.; Thappa, R.K.; Tikku, K.; Chand, D. and Saxena, B.P. 1997. Control of stored product moths and beetles by suboptimum temperatures. *Tropical Sci.*, 37 : 28-34.

Singh, S. 2000. Qualitative and quantitative pest status and eco-friendly management of *Callosobruchus chinensis* Linn. on different pulses. Ph.D. (Ag.) Thesis Submitted to Rajasthan Agricultural University, Bikaner.

Singh, S. and Sharma, G. 2003. Preference of *Callosobruchus chinensis* on pea varieties. *I. J. Ent.,* 65 (2): 277-280

Srivastava, P.K.; Singh, K.; Tomer, R.P.S.; Girish, G.K. and Krishnamurthy, K. 1971. Storage losses in cooperative seed stores in UP. *Bull Grain Tech.*, 9(3): 177-185

Srivastava, P.K.; Tripathi, B.B.; Girish, G.K. and Krishnamurthy, K. 1973. Code of practice for storage of paddy in villages. *Bull Grain Tech.*, 11(3): 206-210

Thompson, R.L. 1915. Sun heat v. weevils. *Rhodesia Agric. J. Salisbury*, 12 : 653-656.

Tiwari, R. and Sharma, V.K. 2002. Resistance to two major stored grain pests in wheat. *I. J. Ent.,* 64 (3): 247-253

White, G.D. 1956. Studies on separation of weevil infested from non-infested wheat by floatation. *U.S.D.A. Circ. Ams.101.* pp. 1-9.

Winterbottom, D.C. 1922. Weevil in wheat and storage of grain in bags. *A record of Australian experience during the war period* (1915-1919). Govt. Printer, Adelaide.

Yadav, R. P.; Bhargava, M.C. and Yadav, J.P. 2002. Studies on the solar radiation as a means of eradicating *Rhizopertha dominica* (Fab.) in wheat. *Indian J. Plant Prot.*, 30 : 26-28.

Index

A

B

C

D

E

F

G

H

I

J

K

L

M

N

O

P

Q

R

U

V

W

Y

Z

FSC
www.fsc.org
MIX
Papier | Fördert gute Waldnutzung
FSC® C083411

Zeitfracht Medien GmbH
Ferdinand-Jühlke-Straße 7
99095 Erfurt, Deutschland
produktsicherheit@kolibri360.de